EAI/Springer Innovations in Communication and Computing

Series Editor

Imrich Chlamtac, European Alliance for Innovation, Ghent, Belgium

Editor's Note

The impact of information technologies is creating a new world yet not fully understood. The extent and speed of economic, life style and social changes already perceived in everyday life is hard to estimate without understanding the technological driving forces behind it. This series presents contributed volumes featuring the latest research and development in the various information engineering technologies that play a key role in this process.

The range of topics, focusing primarily on communications and computing engineering include, but are not limited to, wireless networks; mobile communication; design and learning; gaming; interaction; e-health and pervasive healthcare; energy management; smart grids; internet of things; cognitive radio networks; computation; cloud computing; ubiquitous connectivity, and in mode general smart living, smart cities, Internet of Things and more. The series publishes a combination of expanded papers selected from hosted and sponsored European Alliance for Innovation (EAI) conferences that present cutting edge, global research as well as provide new perspectives on traditional related engineering fields. This content, complemented with open calls for contribution of book titles and individual chapters, together maintain Springer's and EAI's high standards of academic excellence. The audience for the books consists of researchers, industry professionals, advanced level students as well as practitioners in related fields of activity include information and communication specialists, security experts, economists, urban planners, doctors, and in general representatives in all those walks of life affected ad contributing to the information revolution.

Indexing: This series is indexed in Scopus, Ei Compendex, and zbMATH.

About EAI

EAI is a grassroots member organization initiated through cooperation between businesses, public, private and government organizations to address the global challenges of Europe's future competitiveness and link the European Research community with its counterparts around the globe. EAI reaches out to hundreds of thousands of individual subscribers on all continents and collaborates with an institutional member base including Fortune 500 companies, government organizations, and educational institutions, provide a free research and innovation platform.

Through its open free membership model EAI promotes a new research and innovation culture based on collaboration, connectivity and recognition of excellence by community.

More information about this series at http://www.springer.com/series/15427

Josephina Antoniou

Quality of Experience and Learning in Information Systems

Incorporating Learning and Ethics into
Characterizations of Quality of Experience

Josephina Antoniou
School of Sciences
University of Central Lancashire
Larnaka, Cyprus

ISSN 2522-8595   ISSN 2522-8609   (electronic)
EAI/Springer Innovations in Communication and Computing
ISBN 978-3-030-52561-3   ISBN 978-3-030-52559-0   (eBook)
https://doi.org/10.1007/978-3-030-52559-0

This Springer imprint is published by the registered company Springer Nature Switzerland AG
The registered company address is: Gewerbestrasse 11, 6330 Cham, Switzerland

Preface

This book undertakes the notion of evaluating the experience of a user, when the user experiences a new technology. We focus mostly on learning scenarios and we observe the experience aspect from a number of perspectives. We investigate how specific changes in the interface itself can affect experience, but also how the performance of the infrastructure and technology can affect this experience. We explore user expectations, when these expectations have to do with ethics such as trust, bias in design and personal freedoms. The reader navigating through the chapters will come to explore different considerations of user experience for designing new technology products, characterising new types of interactions between users and technology, and understanding how experience is affected by the context of interaction, e.g. learning scenarios.

Well-known notions of experience are presented and discussed, such as user experience, often abbreviated as UX, and user quality of experience, often abbreviated as QoE. We come to the conclusion that a truly positive experience for the technology needs to incorporate aspects of both UX and QoE. These concepts are investigated in different chapters through simple examples and models. In order to address the quantification of experience there needs to be a consideration of how the user experience can be affected by a number of factors. Some of these factors are the contextual environment, the system setup and the network efficiency, but also the interface aesthetics, the software mechanisms, the user behaviour and, in some of the scenarios, the ease with which learning can take place.

Terms such as QoE or UX are coined to refer to the different ways in which user response to a technology is captured. QoE and UX is of interest to us throughout this book, which aims to approach these terms by looking at motivation and decision making that different scenarios, including learning-specific scenarios, may inspire. QoE is an emerging paradigm that puts the user in the centre of the design and evaluation processes of new innovative technological products while still focusing on the quality of the technology itself. UX is an approach that encompasses aspects that deal with the interaction of the user with a particular product or interface. This is mostly focused on user requirements that should be achieved through a satisfactory, simple and aesthetically pleasing solution. Interface design is part of but not all that UX aims to address and evaluate. Changing and emerging technologies motivate us to combine these notions and investigate the experience of the user, as this experience relates to these technologies and to specific scenarios.

Investigating the user experience implies that the user is an active, decision-making entity in the investigated interaction models. The user in different scenarios is required to make decisions based on both complete and incomplete information. The information that the user lacks access to often has to do with another entity's decisions and actions or with matters of technology performance, such as network traffic during the usage of a 5G network, or information about co-existing virtual machines during the usage of a cloud platform.

In fact, user experience can be controversial. On the one hand, interacting with the emerging new technology may turn out to be an easier and more comfortable experience than other alternatives. On the other hand, it can create a sense of lack of control in terms of privacy or personal data protection and other ethical aspects that are important for a satisfying user experience. The increasing deployment of 5G, IoT, Cloud and AI technologies can be either a catalyst for improved user experience, or the opposite, depending on the development and usage approaches. We are not concerned with the overall quantification of user experience but, instead, we aim to offer a closer look into a selected number of interactions and decision-making scenarios, especially in situations where learning or education aspects exist.

The book uses examples of learning and ethics, in several of the investigated scenarios, as well as the way in which these example relate to technology and to the user experience. This approach is chosen in order to investigate relevant decision-making scenarios that may arise with the increasing use of the emerging technologies outlined in the book.

Chapter 1 presents an overview of emerging technologies, focusing on factors and parameters that will be revisited during the following chapters. Emerging technologies are becoming popular, affordable and pushing for a technological revolution in many disciplines. Innovative technology solutions are proposed across the board and increasing numbers and types of users are becoming dependent on them. This revolution has managed to find its way to educational platforms, applications, and both physical and virtual learning tools. The challenge with such educational innovations is that the growth and change in the education practices has not been as rapid as the growth and change in the technology field. Therefore, the user participating in the merging of these two fields to create new solutions may find it challenging in terms of a comfortable experience.

Chapter 2 investigates ways of adopting new educational technologies as part of learning in formal, informal and non-formal educational environments. Towards that goal, there is a need to revisit digital pedagogies to align with new technological capabilities and new user needs. In fact, digital pedagogies need to consider the idea of integrating technology in education as more than just a tool for learning but as an active part of the pedagogy, especially as the need for distance learning delivery and virtual learning environments is increasing. In addition to the idea of digital pedagogies, which are part of more formal environments, concepts such as digital citizenship, whole-school and whole-child approaches are presented to emphasise the widespread need for improved experience with education when it comes to any form of learning (informal, non-formal).

Chapter 3 explores such new technological learning environments and focuses on distance learning platforms. The chapter considers the student preferences as a factor that can affect user experience metrics. Student preferences can be contradicting in group learning environments, and this can be even more challenging in distance learning environments. The chapter explores student preferences as reflected by individual learning styles und used game theoretic models to investigate the interaction between a teacher and the group of students as well as between the students themselves. The chapter tackles user experience and how this is affected by individual learning characteristics and preferences of the students, but also explores the teacher's goal of satisfying those student preferences. Given that learning preferences cannot be completely revealed to the teacher, then ways of improving student experience must be investigated. In addition to the teacher perspective, the chapter focuses on the dynamics between the students themselves during some simple interactions where information processing preferences are used as the parameter that affects students' overall positive experience.

Chapter 4 shifts focus to the technology and discusses the use of Cloud Computing and how this is a trend that has many positive aspects for its users. However, the idea of trust comes into play because Cloud Computing technology is a very abstract concept for many users, and details of its operation are often not revealed. The chapter discusses the technology overall, but quickly shifts focus to a specific example and explores the interaction between a cloud provider and a cloud user from the perspective of trust. Trust between the two interacting parties can affect the user's positive experience. This is even a bigger factor when considering a recurring interaction or an interaction that is continuous over time. The chapter focuses mainly on the perspective of the user. The cloud provider is also significant in the discussion as well as the elements of chance and overall policy. The overall aim is to discuss the motivation for an interaction that will provide a positive experience for the user and a positive payoff for the cloud service provider, as the interaction progresses in time.

Chapter 5 discusses the interface aspect and explores the concept of user experience design and how this design is affected by emotions evoked by interface aesthetics and interface functionality. Concepts such as usability and sociability are presented, as well as challenges in decision-making that the designers face. The selected example deals with digital interfaces that allow for interactions between different users, and the design considers the element of user-to-user communication as part of the experience. The type of digital interface explored in the example is one used for mentoring, i.e. it allows for knowledge exchange between the different types of users. The design of the interface can play an important role in improving the user experience. When referring to experience in relation to a specific interface, we generally refer to elements such as interface usability and design features. To achieve improved UX, designers need to consider elements of user experience as part of the design process, and the overall process is referred to as user experience design.

Chapter 6 addresses emerging AI technologies, but not from a technical aspect. It explores the need for ethics to be integrated with the technical design aspects during the development process of AI products. In order for designers and developers to be able to consider and include ethics in to this process, they have to learn ethics as part of their acquired skills. Therefore, the chapter deals with the teaching and learning aspects of ethics for emerging new technologies, especially AI. It is important to recognise that there are ethical issues associated with the development and use of such technologies. Furthermore, the chapter recognises the trend of an increase in the development and use of AI across many application areas, and ethics must be considered. Moving forward, necessary learning and teaching practices need to be integrated into software development courses. AI is not a standalone technology but a technology that is integrated with society and people and needs to be understood in such an inter-disciplinary manner. The assumption that such courses exist and developers can decide to integrate them into ongoing professional development goals is what the chapter scenario addresses, in an attempt to demonstrate that expectation of ethical design by users can be reflected into resulting user experience from usage of the specific technologies.

Chapter 7 returns the focus to the technology itself. It explores how technology can approach resource management in a user-centred manner. Specifically the chapter takes a look at a 5G resource management scenario to provide a network perspective on the user experience, more as a QoE evaluation in this case. It presents the 5G mechanism of network slicing, which is planned by 5G as a mechanism to address user QoE. Given that network slicing is a wanted solution to support user QoE in 5G networks, specific aspects are explored such as the need to limit the amount of slices in order to avoid a resource scarcity issue and the need to deal with groups of applications assigned to each slice. Groups of applications assigned to a specific slice have similar resource requirements and corresponding quality requirements, but resource management mechanisms still need to address the needs of individual services or applications. The challenge that arises from this grouping of services is that in such a scenario, each service or application has its own specific requirements in addition to the main target requirement for the group, which characterises any specific slice. The scenario presents a scenario that deals with having two applications sharing a slice and having some different requirements. The simplicity of the example gives us the opportunity to discuss some important concepts related to the evaluation of system efficiency in relation to individual user decision-making and how selfish behaviour can affect a system's social good.

Overall, the chapters follow a similar structure. Other than the first two chapters, which focus extensively on technology and learning background knowledge necessary before moving forward, the rest of the chapters follow the following structure.

They, firstly, present the technology and necessary theory or background to understand the scenarios. Then, the scenarios are presented, followed by a mathematical model of the scenario in order to give the reader a more detailed view of decisions and consequences for the entities interacting in each scenario. The chapters finish with a brief discussion of the chapter conclusions.

Larnaka, Cyprus   Josephina Antoniou

Contents

Chapter 1
Quality of Experience and Emerging Technologies: Considering Features of 5G, IoT, Cloud and AI

1.1 Introduction

The increasing popularity of emerging smart technological systems, such as fifth generation (5G) communications, the Internet of things (IoT), cloud infrastructures, cloud computing services and artificial intelligence (AI) applications, highlights the need for consideration of the user experience in terms of quality of the technology used, and in terms of the ease and satisfaction of user interaction with this technology. Terms such as quality of experience (QoE) or user experience (UX) are coined to refer to the different ways in which user response to a technology is captured. QoE and UX will be of interest to us throughout this book, which aims to approach these terms by looking at motivation, and decision-making that different scenarios may inspire.

Quality of experience (QoE) is an emerging paradigm that puts the user in the centre of the design and evaluation processes of new innovative technological products, while still focusing on the quality of the technology itself. User experience (UX) is an approach that encompasses aspects that deal with the interaction of the user with a particular product or interface. This is mostly focused on user requirements that should be achieved through a satisfactory, simple and aesthetically pleasing solution. Interface design is part of UX, but not all that UX aims to address and evaluate. Changing and emerging technologies motivate us to take another look at the experience of the user, as this experience relates to these technologies and to specific scenarios. In terms of scenarios, we will focus on education, learning and related digital interfaces and environments. Chapter 2 presents an overview of the aforementioned aspects. This chapter elaborates on the technologies themselves.

© Springer Nature Switzerland AG 2021
J. Antoniou, *Quality of Experience and Learning in Information Systems*,
EAI/Springer Innovations in Communication and Computing,
https://doi.org/10.1007/978-3-030-52559-0_1

1.2 Characteristics of 5G

The emerging communication technology, referred to as 5G, is a new design for cellular networks so that they can provide, among other applications, high-capacity, low-latency communication for vehicles in highly mobile environments. The new technology, as supportive of smart vehicular networks, is expected to satisfy the requirements laid out for the intelligent transport system (ITS) [13]. In such an environment, the vehicles themselves have a dedicated communication unit, with passengers each having at least one mobile device. The collection of sensors on the vehicles and the activation through the IoT technology is an example of how inter-linked these emerging technologies are. Thus, the 5G vehicular network is closely coupled with intelligent traffic lights and other intelligent city infrastructure, which may include a huge number of sensors and other devices connected with the IoT [2].

To better understand 5G, we will present, in short, four of the main components of this technology that enable it to support new and challenging application scenarios.

The 5G support for mobile edge computing is the first component. Mobile edge computing is a way for bringing services to the most suitable network location. Discovery, access and advertisement of relevant services are defined [1]. This is similar to an offloading service for vehicular networks such that, with sufficient computing resources, mobile edge computing can outperform in-vehicle computing capabilities in the execution of computation-intensive tasks and in delivering lower execution latency.

A variation of this approach is vehicular fog computing [21], which employs underused vehicles as the infrastructure for task computation and communication. In general, the idea of edge computing, used in the above examples, is about moving closer to the edge of the network, and of processing and analysing data in servers closer to the applications, and hence the users, that the applications serve.

A very interesting second component of 5G networks is that of network slicing. Network slicing addresses the key challenge in 5G, which is the management of all the available heterogeneous access networks in terms of service requirements. Network slicing aims to ease resource management by logically separating the net-works. For instance, transport safety applications can be specified as a network slice that requires lower latency and highly reliable periodic message transmission. Other logical network slices can be designated for many other demanding applications. The idea is that similar requirements' applications are placed in the same slice so as to satisfy their QoS requirements and to enhance user experience. Later in the book, we will further explore network slicing and user experience, by considering pro-posed slicing by standards organisations and how each slice needs further inspec-tion to ensure user QoE.

Another 5G component is that of a software-defined network, which specifies a layered network structure, where the control layer provides efficient centralised management over the underlying infrastructure through software modules. Although, software-based, session control mechanisms have been used in previous generation of communication technologies [4], routing based on a software-defined network in 5G is advantageous. This is evident when looking at the packet delivering ratio compared with traditional routing protocols. This is especially advantageous for highly mobile networks, such as vehicular networks [15]. Advantages of such approaches often include reducing latency and improving resource utilisation rate.

The fourth component is more related to services, and it is that of network functions virtualisation, which consists of the network functions that form a slice or a service to be implemented in a virtualised way [8]. This component, in combination with software-defined radio, can allow different tenants, in multi-tenant systems, to share the same general-purpose hardware and to build fully decoupled end-to-end networks on top of a common, shared infrastructure.

Given the proposed technology components that characterise 5G, we need to consider the effect of this new technology on the experience of the 5G user. Ensuring QoE from a technology perspective for 5G users becomes very challenging, in spite of the technological advances that 5G demonstrates. The 5G user is no longer the owner of a just simple smartphone device, but may be a user of IoT or a vehicular network, in which case the "things" or "vehicles" that interoperate with the user (and with other "things" and "vehicles"). In addition, the increased number of users and the variation of the types of users, because of the interoperability with other technologies (e.g. IoT), results in a variety of new applications, a variety of subjective requirements and poses challenges in being able to understand and satisfy QoE across the board.

Investigating different scenarios and dealing with case-by-case requirements is an approach that is expected to slowly bring results into questions such as what measurements are needed by the network operators and how should these be interpreted to be able to improve user QoE. The book approaches a variety of scenarios in this manner, by addressing simple interactions and user requirements, to demonstrate the "experience-focused" view of technology.

1.3 Characteristics of IoT

With the deployment of 5G networks, end-users will be able to interact with multiple IoT applications, leveraging the benefits offered by such smart environments [19]. It is expected that the success of 5G deployment will increase the number of cellular IoT connections exponentially [18], with Ericsson forecasting up to 3.5 billion by 2023 [6]. Research in the domains of IoT and smart environment aims to contribute to the generation of new ideas leading up to the development of new technological enhancements that support products with an improved QoE. The idea of responsible research and innovation (RRI) is becoming increasingly important as well, given that responsible development will be reflected onto the quality of the developed products and lead to responsible use of such technological products. The societal and environmental benefits from responsible development and use will have consequent benefits on the users' experience.

Considering the new innovative applications are being implemented, IoT is no longer a technological buzz word; rather a revolution, which is well underway. As smart devices and sensors are becoming more and more ubiquitous and weaved within our environment, IoT applications can cover (almost) all aspects of our life. From industrial and manufacturing domain to agriculture and farming, healthcare, logistics and transport to smart homes and cities, IoT applications are the emerging technological enablers. The worldwide IoT market is expected to reach *41.6 billion connected IoT devices, or "things", generating 79.4 zettabytes (ZB) of data in 2025*, according to IDC [11]. Extensive research has been conducted at various aspects of IoT including enabling technologies, standards, architecture and models and security and privacy concerns within various domains. However, IoT service adoption from the user's perspective has not received significant attention [9]. This makes this an interesting area of investigation for user behaviour and experience.

From an end-user perspective, IoT applications can vary from simple personal activity tracking gadgets to high-end connected cars. A user's intention to use an IoT application is based on perceived benefits, ease of use, reliability and security, to summarise the overall QoE. QoE can be an important way to measure how well an automated system performs, as well as its overall acceptability and usability. The QoE measurement considers the complete end-to-end system; the user, the end-device(s), underlying network infrastructure(s), services and application/content.

Traditionally, emphasis was given to the device and the network characteristics and measurements to quantify quality of service (QoS), which works well if the recipient of an IoT application is another computer system or application. However, QoS-based evaluation of quality fails to include the end-user experience and perception of the IoT application. In order to achieve that, we need to evaluate the quality within the context of QoE measurements.

For QoE evaluation, it is imperative to formulate a clear understanding of what is QoE and how it is affected by the underlying system. QoE requirements can vary based on the IoT application domain; furthermore, QoE requirements can even vary among IoT applications belonging to the same IoT domain. Generically, QoE is defined as the degree of delight or annoyance of the users of an application or service, by the International Telecommunication Union (ITU) [12]. The European Telecommunications Standards Institute (ETSI) extends the definition to include both technical parameters and usage variables and measures both, the process and outcome of communication. It defines QoE as a measure of user performance based on both objective and subjective psychological measures of using ICT service or product [7]. The European Network on QoE in Multimedia Systems and Services, Qualinet, defines it, similarly to the ITU definition, as *the degree of delight or annoyance of the user of an application or service* [5]. It results from the fulfilment of a user's expectations with respect to the utility and of the enjoyment received from the application or service in light of the user's personality and current state.

Extracting from these definitions, ITU has proposed two methods to measure user experience; (1) subjective QoE assessment, typically based on mean opinion score (MOS) of a service according to user perception and (2) objective QoE assessment, typically involving quality of service (QoS) parameters like latency, traffic volume density, reliability, cost, etc. These sets of guidelines align with the view of this book to consider QoE and UX as two different approaches of evaluating user experience.

1.4 Characteristics of the Cloud

IoT is also a supporting technology for Big Data, since it is one of the sources of a huge amount of data collection that can be manipulated by such platforms as cloud computing platforms and can be used to power emerging applications that use Big Data to generate intelligence, such as artificial intelligence (AI) and machine learning applications.

Cloud computing platforms are becoming more and more popular as they offer several advantages to users, especially to business owners. Advantages include lower costs for hardware, elasticity of resources and the ability to plug into many available cloud services. This seemingly should improve user experience, as the user may experience more and better quality (due to the availability of resources) services.

In general, cloud computing platforms offer four different types of services to cloud users: the Infrastructure-as-a-Service (IaaS), the Platform-as-a-Service (PaaS), the Function-as-a-Service (FaaS) and the Software-as-a-Service (SaaS). Specifically, for IaaS, the user can create virtual machines from scratch, by selecting versions of operating systems made available by the cloud provider. Once that is configured, the user is responsible to set-up and manage the networking, i.e. configurations that control communication with the virtual machines. For PaaS, the user can configure the environment instance that would allow for the development of new software. The container environment allows for the user to control, any client – server communication for the developed software, but does not allow access to the operating system. For FaaS, similar access control is given to the user, except the ability to edit any configuration related to the server side of the software development, because FaaS only allows for client-side implementation. For SaaS, the user only has access to specific application and cannot change any system configurations.

However, there are still some risks to consider that may affect user quality of experience. By using the cloud, the users give up control of their data, which is stored in an unknown location, and possibly on the same machine with other users' data. Such aspects of trust and service agreements between cloud providers and users will be explored later in the book, as we view trust to be key to a satisfying user experience in interacting with cloud computing technologies.

1.5   Characteristics of AI Applications

Popular cloud platforms often offer AI and machine learning applications as part of their service package. Such intelligent applications, as smart transcription of audio files, smart analysis of photographs, emotion detection in text or audio, etc., gain more and more popularity among cloud users. Other applications such as data modelling and machine learning for prediction purposes, or identification of trends, are also quite popular with business owners that user cloud computing platforms as part of their business infrastructure.

AI will live at every edge in the hybrid clouds, multiclouds and mesh networks of the future. Already, prominent AI platform vendors make significant investments in 5G-based services for mobility, IoT and other edge environments [14].

Machine learning is one of the AI flavours that intelligent applications are based on. AI as an area of study, in particular through the use of machine learning mechanisms, often attempts to simulate human intelligence characteristics in machines and software, such as learning and decision-making. In fact, the increasingly large amount of data available over the Internet makes the *learning* part of AI easier.

AI in combination with Big Data can also contribute to research as they can support the various aspects of the scientific process. With respect to making observations, descriptive analytics on Big Data [20] focus on classification and clustering of data and their visualisation, which can allow users to visualise a holistic picture based on relationships predetermined by the users themselves. This reduces complexity, allowing users, in particular, researchers, to comprehend the big picture and develop hypotheses. Machine learning techniques such as deep learning [16] can undertake exploratory research by analysing Big Data, recognising and extracting patterns from observation and analysis and eventually generate new hypotheses.

Furthermore, AI and machine learning techniques can be used to augment and increase the efficiency of some labour-intensive research and innovation activities [17] such as literature search, data collection and clustering [10]. Finally, in terms of interpreting the results of a research and innovation project, AI can be used as an autonomous system, which can make its own decisions relevant to learning but also developing judgments about, a specific project or experiment.

The questions arising from the use of AI in research are related to user experience through the consideration of ethics. Our learning approach becomes very relevant to AI and ethics, since there arises the need to educate AI developers and users about ethical issues. Because of the automated learning and decision-making, the developers of AI systems must consider ethics within the design to ensure that any automated decisions consider ethics and are responsible. Similarly, when such technologies are used, especially to support innovation in society, then ethical use is critical. Thus, there is a need for a moral code to support the decision-making processes of the developers, the software and the users of AI systems, and the book will investigate a potential approach towards teaching and learning of ethics for AI.

1.6 Discussion

In summary, amongst many new technological options, we may witness how technology has come together to enable new opportunities, applications and services. Big Data, generated in part by the IoT, can be stored in enormous data centres, that can be part of distributed infrastructures like the Cloud, and can be analysed by new machine learning and AI algorithms, to power demanding new mobile applications that are supporting by a next generation enhanced mobile network, the 5G network.

The idea of everything being connected together is becoming a reality and the user is in the middle of it all. The user experience is important to understand, analyse and improve where possible, whether affected by visceral or behavioural reactions, by the system performance itself or even by ethical concerns that may improve or not the quality of this user experience.

Considering the large number of new challenges when it comes to emerging technologies and new applications, we explore what we view as in the core of all these challenges: to keep the user in the centre of any technological decisions, which should be ethical and improving the overall user experience. Some of these decisions may have been viewed within the context of older technologies [3], such as making a selection within a technological environment, but the book will follow in detail the decisions for the specific scenarios and environments in focus, with a recurring interest for learning- and ethics-related scenarios.

The increasing deployment of 5G, IoT, Cloud and AI technologies can be a catalyst for improved user experience, or not, depending on the development and usage approaches. We are not concerned with the overall quantification of user experience but, instead, we aim to offer a closer look into a selected number of interactions and decision-making scenarios, especially in situations where learning, education or ethics-related aspects exist.

References

1. Ahmed A, Ahmed E (2016) Survey on mobile edge computing. In: 10th International Conference on Intelligent Systems and Control (ISCO), pp 1–8
2. Al-Fuqaha A, Guizani M, Mohammadi M, Aledhari M, Ayyash M (2015) Internet of things: a survey on enabling technologies, protocols and applications. IEEE Commun Surv Tutor 17(4):2347–2376. https://doi.org/10.1109/COMST.2015.2444095
3. Antoniou J, Pitsillides A (2006) Radio access network selection scheme in next generation heterogeneous environments. In: Proceedings of the IST Mobile Summit
4. Antoniou J, Christophorou C, Neto A, Sargento S, Pinto F, Carapeto N, Mota T, Simoes J, Pitsillides A (2012) Session and network support for autonomous context-aware, multiparty communications in heterogeneous mobile systems. In: Emergent trends in personal, mobile, and handheld computing technologies. IGI Global, Hershey, pp 264–285
5. Brunnström K, Beker SA, de Moor K, Dooms A, Egger S et al (2013) Qualinet white paper on definitions of quality of experience. hal-00977812
6. Ericsson (2018) Ericsson mobility report. Retrieved from: https://www.ericsson.com

7. European Telecommunication Standard Institute (2010) ETSI TR 102 643: Human Factors (HF) Quality of Experience (QoE) requirements for real-time communication services [Online]. Available: http://www.etsi.org

8. European Telecommunication Standard Institute (2014) Network Functions Virtualization (NFV); Architectural Framework, ETSI GS NFV 002 v1.2.1 (2014-12). Retrieved from: http://www.etsi.org

9. Floris A, Atzori L (2015) Quality of experience in the multimedia internet of things: definition and practical use-cases. In: 2015 IEEE International Conference on Communication Workshop (ICCW), London, pp 1747–1752

10. Gil Y, Garijo D, Ratnakar V, Mayani R (2016) Automated hypothesis testing with large scientific data repositories. In: Proceedings of the Fourth Annual Conference on Advances in Cognitive Systems (ACS), pp 1–6

11. IDC (2019) The growth in connected IoT devices is expected to generate 79.4ZB of data in 2025, according to a new IDC forecast. Retrieved from: https://www.idc.com

12. International Telecommunication Union (2017) ITU-T P.10/G.100 Vocabulary for performance, quality of service and quality of experience [Online]. Available: http://www.itu.int

13. Karagiannis G, Altintas O, Ekici E, Heijenk G, Jarupan B, Lin K, Weil T (2011) Vehicular networking: a survey and tutorial on requirements, architectures, challenges, standards and solutions. IEEE Commun Surv Tutor 13(4):584–616

14. Kobielus J (2019) How 5G will Serve AI and Vice Versa. Retrieved from: https://www.datanami.com

15. Ku I, Lu Y, Gerla M, Gomes RL, Ongaro F, Cerqueira E (2014) Towards software-defined VANET: architecture and services. In: 13th Annual Mediterranean Ad Hoc Networking Workshop (MED-HOC-NET), pp 103–110

16. LeCun Y, Bengio Y, Hinton G (2015) Deep learning. Nature 521:436–444. https://doi.org/10.1038/nature14539

17. MacCallum KJ (1990) Conceptual design environments-research directions and issues. In: IEE Colloquium on Strategic Research Issues in AI in Engineering, London, pp 1/1–1/4

18. Palattella MR et al (2016) Internet of things in the 5G era: enablers architecture and business models. IEEE J Sel Areas Commun 34(3):510–527

19. Shahzad M, Antoniou J (2019) Quality of user experience in 5G-VANET. In: IEEE 24th International Workshop on Computer Aided Modeling and Design of Communication Links and Networks (CAMAD), pp 1–6

20. Sun Z, Zou H, Strang K (2015) Big data analytics as a service for business intelligence. In: Janssen M et al (eds) Open and big data management and innovation. I3E 2015. Lecture notes in computer science, vol 9373. Springer, Cham

21. Xiao Y, Zhu C (2017) Vehicular fog computing: vision and challenges. IEEE Comput Soc Digit Libr 1:6–9. https://doi.org/10.1109/percomw.2017.7917508

Chapter 2
Adopting New Educational Technologies and the Need for Digital Pedagogies

2.1 Introduction

The book uses the example of education, in several of the investigated scenarios, as it relates to technology and to the user experience. This approach is chosen in order to investigate relevant decision-making scenarios that may arise with the increasing use of the emerging technologies outlined in Chap. 1.

In addition to emerging technologies, other factors, such as the popularity and affordability of personal computers, have sparked a technological revolution, which has not only resulted in innovative technology solutions but has also found its way to educational platforms, applications and both physical and virtual learning tools. This has also influenced the rate at which new educational technologies affect learning and teaching experiences. It has been observed that while citizens as part of modern societies have embraced technological advancements, educational players, both practitioners and policy-makers, have kept well behind in embracing technological change.

Oftentimes, the examples of educators that take advantage of new technologies still offer the perspective of technology in education as a useful tool and nothing more. Nevertheless, technology can serve a higher purpose than a simple tool to facilitate traditional education. Specific applications and technological infrastructures, collectively referred to as educational technologies, can often be used to improve teaching and learning. To achieve such significant change, we may consider the idea of digital pedagogies to complement existing pedagogical methods, for learning environments where educational technologies can be introduced.

© Springer Nature Switzerland AG 2021
J. Antoniou, *Quality of Experience and Learning in Information Systems*,
EAI/Springer Innovations in Communication and Computing,
https://doi.org/10.1007/978-3-030-52559-0_2

2.2   Characteristics of Digital Pedagogies

A digital pedagogy is an attempt to use technology to change teaching and learning in a variety of ways. A digital pedagogy is not merely concerned about replacing the traditional tools like blackboards and textbooks with iPads and websites, which would be feeding in the "education as a tool" perspective. Overall, a digital pedagogy is more concerned with how, for an education ecosystem, learning and literacy are fostered by the use of digital and electronic media [4].

The experience of the learner in such environments must be revisited. Interface aesthetics or efficient communication channels are no longer enough for a positive experience but instead the evaluation of learning becomes the main concern. Changing the medium of delivery to a digital interface does not necessarily complicate or simplify the learning. It is, however, important that the teacher plans the delivery of the learning with the consideration of old and new learning challenges that the learners have to face, that range from satisfying personal learning preferences and inclinations, to receiving the content of the lesson through digital interfaces that are easy to navigate and understand.

Another important aspect to consider is the quality of e-learning software and consequent digital pedagogies. Published research recognises the complicated task of characterising learning. An article published in 2016 [28] produced a set of heuristics that consider a number of learning theories. According to the authors, the pedagogical heuristics consider *theories related to technology and our current digital society*, and overall results of the study show that *e-learning can offer educational benefit*. The authors have attempted to use multimodal learning as part of their e-learning software, i.e. to represent the educational material through approaches that cater to various learning preferences, such as diagrams, videos, audio and textual representations, images as well as activities. We will explore such preferences in subsequent chapter. We note, however, that a multimodal approach to e-learning appears to be a successful pedagogical approach.

Digital pedagogical strategies need to be developed to support such online learning. A digital pedagogy is a set of tools for learning that explore the use of digital resources, their effectiveness and their impact. Such strategies also consider improving the teaching by the use of digital technology. Overall, digital strategies in education have the potential to make education more accessible and inclusive and support lifelong learning. We will approach the lifelong learning opportunities in a subsequent chapter.

2.3   Characteristics of Digital Citizenship

Teaching and learning does not only happen in classrooms, whether these are physical or virtual. Teaching and learning happens through everyday living, especially through the continuous interaction with technology in smart environments using our smart devices. Ubiquitous access to technology has revolutionised every aspect of our lives and transformed our communities into digitally connected citizens. Digital citizenship refers to the new codes of conduct citizens must adapt to, in order to be safe, responsible and respectful participants in today's digitally connected culture.

Responsibility is a major consideration of both designing and using digital pedagogies, participating in a digitally enriched society and using corresponding digital platforms and interfaces. In addition to using technology, responsibility is a key parameter of developing technology, especially when developing new technological products for education. Thus, we need to consider responsibility from the research phase of the project development cycle.

2.4   Characteristics of Responsible Research and Innovation

The generation of new ideas, leading up to the development of new technological enhancements, is a process that must eventually support responsible technological products that can support an improved user experience. The idea of responsibility in this case represents technology that is aware of societal needs. Responsible research and innovation processes are reflected onto the quality of the developed products with consequent benefits on the products' users, the environment and society overall. On the contrary, *poor research practices have detrimental consequences on the quality of research outputs* [19]. European Commission, acknowledging the rapidly growing significance of RRI, called for further research (e.g. through FP7,[1] H2020[2] calls for project bids) to get a more comprehensive understanding on how responsible research and innovation can be promoted.

Responsible research and innovation processes can improve research on digital pedagogies and educational technologies for improved QoE and UX. The rationale behind choosing to focus on user experience from the research phase is so that the designers may consider, from the beginning, the input from key stakeholders into the technological products. Considering input from potential product users leads to a responsible implementation and usage of any given technological product. According to the European Commission's definition of responsible research and innovation from the Responsible Research and Innovation (RRI) report [16] published under the Science in Society initiative:

> RRI means that societal actors work together during the whole research and innovation process in order to better align both the process and its outcomes with the values, needs and expectations of European Society [...] an ambitious challenge for the creation of a Research and Innovation policy driven by the needs of society and engaging all societal actors via inclusive participatory approaches.

RRI is a concept promoted, especially within the Horizon 2020 programme, to address the application of research and innovation in industry. Through the eighth

---

[1] The Seventh Framework Programme by the European Commission.

[2] Horizon 2020 is an EU research and innovation programme available from 2014 to 2020.

European Framework Programme, widely known as Horizon 2020 (or H2020), the effort to align research and industry is used as a means to address societal challenges through research and innovation. The main point of contention for the implementation of RRI seems to be the trade-off between *the race to stay competitive in a rapidly changing world* and *the need to maintain public trust* [22].

Implementing RRI *requires collaboration of various stakeholders in order to find sustainable solutions based on ethical acceptability, sustainability and societal desirability* [22], but the variety of components that must be considered in the RRI implementation process can be overwhelming [1]. In fact, RRI components are often referred to as *RRI Core Pillars* in relevant projects, such as by the Horizon 2020 project [24], which made use of the idea of RRI Core Pillars, i.e. *engagement, gender equality, science education, ethics, open access and governance.*

Nevertheless, *attention is not routinely paid to the ethical and social implications* of research and innovation activities in industry, as issues of business ethics for corporate activities have been extensively studied in research [6]. By investigating the RRI core pillars, it is evident that they cannot be useful in all possible corporate scenarios in the same way, especially where applications in education are concerned. It has been shown, however, that while profit-oriented incentives are certainly viable, an organisation's contribution to solving societal challenges is also an overarching benefit. Corporate innovation management has been addressed within many corporate strategies, to provide a solution for the challenge of workforce being trapped within a certain routine without ever thinking outside the box [23]. RRI encourages an approach towards innovation where *societal actors and innovators become mutually responsive to each other with a view to the acceptability, sustainability and societal desirability of the innovation process* [25].

As the effort to intertwine scientific excellence and society, through the implementation of responsible practices, has been increasing, so have the observed challenges [21]. While there have been many studies of RRI throughout projects funded by European Framework Programmes, even prior to the launch of Horizon 2020, interdisciplinary research has attempted to explore in depth the main challenges towards the implementation of RRI for improving user experience and learning.

2.5 Characteristics of Responsible Digital Citizenship Education

Learning can be achieved in several ways. Traditionally, formal education uses systematic ways of achieving academic objectives in structured environments such as schools and classrooms. The concept of informal education refers to the learning that happens outside the classroom, such as learning life skills, for example values and traditions. We are concerned with structured or systematic learning in subsequent chapter; however, this can still happen outside the classroom in a non-formal way, which is usually more flexible, more practical and focused on the learner.

Learning outside the classroom can serve the additional purpose of supporting citizens in their home and professional lives. A European Digital Competence Framework, *DigComp* [10], was formulated to help the citizens with self-evaluation, setting their learning goals, identifying training opportunities and facilitating job search. The framework lists 21 competences with 8 proficiency levels. In addition, it provides an aid to policy-makers by offering an indicator for digital skills.

In fact, the Council of Europe emphasises the need to educate citizens of this digital world, especially the young population, regarding their rights and well-being online as well as offline and suggests that in addition to policies to protect children online, it is important to empower them to become healthy and responsible online citizens [7]. The Digital Citizenship Education initiative is thus crucial and involves active discussion revolving around the Internet safety and security, cyberbullying, relationships and communities, digital identity, information literacy and other areas of concern. It is important that citizens are encouraged to develop their online proficiency, engagement and creativity as well as an awareness of the legal implications of their online activity.[3] Digital platforms that support mentoring schemes, trainings and knowledge exchange activities are important in alleviating the effects of the digital divide in education across different regions, different ages and different professions.

It is, therefore, important to strengthen active citizenship, and this can be supported through digital educational means. Digital education happens not only in a classroom but also online, for example through social media. For example, responsible citizens learn to identify false information or know how to filter junk messages. A relevant project, called *Detect* [9], aims to develop training for teachers in order for them to familiarise themselves with relevant tools and strategies that manipulate opinion and behaviour. Once the teachers acquire this knowledge, they can transfer the acquired knowledge to their students.

New digital skills and new learning environments can foster the empowerment of children, and learners in general, through education. By enriching societal aspects with technology, the children can apply their new digital competencies in their everyday life, and, they can become active participants in this new digital society.

Digital Citizenship Education can explore different domains. Domains that a digital citizen can develop new skills through educational technology may include policy-making, infrastructure design or usage, new strategies for business or education, environmental and sustainability support, etc. The learning needs to be supported by setting strong foundations in terms of values, behaviours and understanding, in addition to the new skills that the learner will develop. This takes us back to the idea of responsibility and creates room for the consideration of ethics, for supporting the society and citizens' quality of life. Ethics will be one of the elements that will concern us in subsequent chapters as it relates to experience.

---

[3] Digital Citizenship: https://www.coe.int/en/web/digital-citizenship-education/digitalcitizenship-and-digital-citizenship-education.

The book will not explore digital citizenship per se, but it will address aspects of this approach as part of the exploration into user experience. So, for the purpose of this exploration, aspects such as digital engagement, digital responsibility and digital participation will be addressed throughout. Digital engagement refers to the competent and positive engagement with digital technologies and data. Digital engagement activities include, but are not limited to, developing, sharing, researching, working, socialising, communicating and, of course, learning. Digital responsibility refers to the consideration of important elements such as ethics, gender, open access, etc. when developing and using technology. Responsibility is a concept that considers values, skills, attitudes, knowledge and critical understanding. Digital participation refers to ways of actively participating in local, national and global communities, at any societal levels, e.g. social, political, cultural, etc. while simultaneously being involved in a lifelong learning (formal, informal, non-formal) to allow the responsible citizen to continuously support ethics and human rights.

Digital Responsible Citizenship in a Connected World (DRC) [13], an ERASMUS+[4] funded project, aims to impart digital pedagogical practices in both, teachers and students across Europe and follow the competences identified in the *DigComp*. The project uses gamified adventure environment to teach primary school students the fundamentals of digital citizenship. The project comprises five phases for targeting digital literacy, specifically: information and digital literacy; communication and collaboration; digital content creation; safety; problem-solving.

Similarly, another project BRIGHTS [5] aims to promote Global Citizenship Education (GCE) in formal and non-formal educational contexts in Europe, with the help of digital storytelling (DS) techniques. All digital stories published on the project's website focus on global citizenship education topics, such as, for instance, sustainable development and lifestyle, social inclusion, cultural diversity, gender equality, peace and human rights, etc.

More projects that make use of this concept have been funded across Europe. For example, project EMPOWER: Empowering Digital Citizenship Through Media Literacy and Critical Thinking [15] targets young people and how they can responsibly consume and create online and social media content. The project aims to deliver a good practices catalogue; developed based on the concept of knowledge sharing and peer learning, a digital citizenship toolkit; comprising online resources for teachers and educators and digital pedagogy courses.

User experience for students needs to be considered within their learning environments. The school environment has often been identified as a key setting for building social, emotional and behavioural outcomes because students spend a substantial amount of time there [18]. The school also provides a socialising context in which students are able to learn a range of life skills, many of which are associated with academic success [27]. Considering learning environments as crucial in evaluating user experience can be extended to all learning environments. As

---

[4] Erasmus+ is an EU programme that supports education, training, youth and sport in Europe.

learning happens in formal, informal and non-formal settings, then the idea of environment is expanded, and the consideration of user experience must often evaluate scenario-specific situations.

2.6   Characteristics of Whole-School and Whole-Child Approaches

A whole-school approach aims to raise quality and standards across the entire school by providing learning opportunities for students in every aspect of their lives [17]. For this approach to be effective, schools need to identify and address the needs of the school community and engage in continuous, cyclical processes for improvement while engaging all the members of the school community (management, teaching and non-teaching staff, learners, parents) and supporting increasing diversity [20].

Several countries have launched national initiatives that adopt a school-wide approach to social and emotional learning, for example KidsMatter Primary [12] and the *Be You* initiative [3] in Australia, the Social and Emotional Aspects of Learning (SEAL) programme in the UK [8], Green School award in Sweden and the ELP-WSU European initiative [14]. KidsMatter Primary is a mental health and well-being initiative for Australian primary schools, which aims to identify and implement whole-school strategies to improve students' mental health and well-being. Similarly, BeYou is an initiative for Australian education that provides educators with knowledge, resources and strategies for helping children and young people achieve good mental health. The *SEAL* programme uses a whole-school approach to achieve effective learning, positive behaviour, regular attendance, staff effectiveness and the emotional well-being of all involved stakeholders. The Green School Award is an initiative that identifies and celebrates sustainability in education. The ELP-WSU initiative aims to study existing whole-school approaches in an attempt to develop guidelines for the design, implementation and management of whole-school projects and communicate them to policy-makers.

A whole-child approach to education is defined by policies, practices and relationships that ensure each child, in each school, in each community, is healthy, safe, engaged, supported and challenged [11]. Within a whole-child approach, questions must be raised about school culture and curriculum; instructional strategies and family engagement; critical thinking and social-emotional wellness [2]. This framework has been used as the scaffold in the development of a range of school improvement processes that ensures that the approach is integrated and systemised into the processes and policies of the school, district and community [26]. The key components are: school climate and culture, curriculum and instruction, community and family, leadership, professional development and capacity and assessment. Although this approach was originally launched in the USA, it has become a widely adopted initiative with countries across the Americas, Europe and Oceania using the approach to refocus their educational systems on local, regional and national levels. A whole-child approach to education is an approach that spans content areas and demands whole-school improvement [26].

2.7   Discussion

This book focuses on quality of user experience in scenarios where emerging technologies play a key role, but also in scenarios where education plays a key role. We have attempted to explore relevant and interesting topics for the learning component of user experience such as digital pedagogies or whole-school and whole-child approaches. We have also explored the importance of responsibility and ethics for technologically enhanced environments, especially within learning scenarios. The subsequent chapters will further investigate selected interactive examples, model them using mathematical tools in order to better understand and analyse them. We have attempted to analyse a number of examples that combine the elements of experience, learning, ethics and technology in the subsequent chapters.

Moreover, the chapter discussed the significance of digital pedagogies to achieve improved student experience and learning environments. The variety of projects, initiatives and publication shows the broad recognition of the urgency of addressing the gap in education that fails to follow the rapid growth of technology. It is important to address the societal effect of any changes made in either education or technology or a combination of both disciplines within the context of educational technologies or digital pedagogies. That may be found within the experience or responsibility components of the scenarios discussed henceforth. The consideration of responsibility, ethics and user experience are aspects that have been hitherto considered key to successful innovations in both the technology and education disciplines. Thus, in the selected scenarios, it is expected that consideration of responsibility, ethics and user experience in the design and use of technology and learning (e.g. digital pedagogies or learning environments) will positively impact the overall outcomes.

Through modelling and analysing the following scenarios, organised in the five remaining book chapters, we aim to develop a clear and systematic understanding of the ways in which technology and learning can be approached such that they result in improved user experience. Such improved experience may be evident from improved utility of users, or from a more responsible and ethical decision-making between the interacting users. The users in the scenarios undertake various roles, e.g. as teachers or learners, mentors or mentees, service providers or service customers and, in some cases, more generically, technology users.

The book plans to provide models of interaction in specific scenarios and analyse them further, in order to extract some conclusions from the presented and analysed models.

References

1. Adams R, Jeanrenaud S, Bessant J, Denyer D, Overy P (2016) Sustainability-oriented innovation: a systematic review. Int J Manag Rev 18(2):180–205
2. ASCD (2020) A whole child approach to education and the Common Core State Standards Initiative [Online]: http://www.ascd.org/ASCD/pdf/siteASCD/policy/CCSS-and-Whole-Child-one-pager.pdf
3. Be You (2020) [Online]: https://www.beyou.edu.au/
4. Blewett C (2016) From traditional pedagogy to digital pedagogy. In: Samuel MA, Dhunpath R, Amin N (eds) Disrupting higher education curriculum. Constructing knowledge: curriculum studies in action. Sense Publishers, Rotterdam. https://doi.org/10.1007/978-94-6300-896-9_16
5. BRIGHTS (2017) [Online]: http://www.brights-project.eu/en/
6. Chatfield K, Borsella E, Mantovani E, Porcari A, Stahl BC (2017) An investigation into risk perception in the ICT industry as a core component of responsible research and innovation. Sustainability 9(8):1424
7. Council of Europe (2019) Digital citizenship education handbook. Council of Europe Publishing, Strasbourg. ISBN 978-92-871-8734-5
8. Department for Children, Schools, and Families (2008) SEAL: Social and Emotional Aspects of Learning, Curriculum resource introductory booklet. Ref: 00258-2008DWO-EN-02
9. Detect (2020) [Online]: https://www.detect-erasmus.eu/en/about-us/
10. DigComp 2.0 (2018) [Online]: https://ec.europa.eu/jrc/en/digcomp/digital-competence-framework
11. DigCompEdu (2019) [Online]: https://ec.europa.eu/jrc/en/digcompedu
12. Dix K, Keeves JP, Slee PT, Lawson MJ, Russell A, Askell-Williams H, Skrzypiec G, Owens L, Spears B (2010) KidsMatter primary evaluation: technical report and user guide. https://research.acer.edu.au/learning_processes/23
13. DRC (2017) Digital responsible citizenship in a connected world [Online]: https://digital-citizenship.org/
14. ELP-WSU (2011) [Online]: https://elp-wsu.ecml.at/
15. EMPOWER (2020) Empowering digital citizenship through media literacy and critical thinking [Online]: http://www.empowerme-project.eu/about/
16. European Commission (2012) Responsible research and innovation: Europe's ability to respond to societal challenges
17. European Commission (2018) A whole school approach to tackling early school leaving Policy messages, Education & Training 2020: Schools Policy. https://ec.europa.eu/education/sites/education/files/document-library-docs/early-school-leaving-group2015-policy-messages_en.pdf
18. Goldberg JM, Sklad M, Elfrink TR et al (2019) Effectiveness of interventions adopting a whole school approach to enhancing social and emotional development: a meta-analysis. Eur J Psychol Educ 34:755–782. https://doi.org/10.1007/s10212-018-0406-9
19. Iordanou K (2019) Involving patients in research? Responsible research and innovation in small- and medium-sized European Health Care Enterprises. Camb Q Healthc Ethics 28(1):144–152
20. Jones SM, Bouffard SM (2012) Social and emotional learning in schools: from programs to strategies: social policy report. Soc Res Child Dev 26(4):3–22
21. L'Astorina A, Fiore MD (2017) A new bet for scientists? Implementing the responsible research and innovation (RRI) approach in the research practices. Beyond Anthropocentrism 5(2):157–174
22. Martinuzzi A, Blok V, Brem A, Stahl B, Schönherr N (2018) Responsible research and innovation in industry – challenges, insights and perspectives. Sustainability 10(3):702

23. Mitra S (2016) Corporate innovation management: lessons learned. One Million by One Million Blog. Available at: https://www.sramanamitra.com/2016/07/22/corporate-innovation-management-lessons-learned/. Accessed 13 Nov 2016

24. Responsible Industry (2014) Responsible Research and Innovation in Business and Industry in the Domain of ICT for Health, Demographic Change and Wellbeing. http://www.responsible-industry.eu/

25. von Schomberg R (2012) Prospects for technology assessment in a framework of responsible research and innovation. In: Technikfolgen abschätzen lehren. VS Verlag für Sozialwissenschaften, Wiesbaden, pp 39–61

26. Slade S, Griffith D (2013) A whole child approach to student success. KEDI J Educ Policy:21–35

27. Taylor RD, Oberle E, Durlak JA, Weissberg RP (2017) Promoting positive youth development through school-based social and emotional learning interventions: a meta-analysis of follow-up effects. Child Dev 88(4):1156–1171. https://doi.org/10.1111/cdev.12864

28. Yiatrou P et al (2016) The synthesis of a unified pedagogy for the design and evaluation of e-learning software for high-school computing. In: Ganzha M, Maciaszek L, Paprzycki M (eds) Proceedings of the 2016 federated conference on computer science and information systems, ACSIS, vol 8. Polskie Towarzystwo Informatyczne/IEEE, Warsaw/Los Alamitos, pp 927–931

Chapter 3
New Technological Learning Environments: Tensions Between Teaching and Learning in Groups and Consideration of Learning Styles to Improve Quality of Student Experience

3.1 Introduction

This chapter tackles user experience and how this is affected by an individual's characteristics in a learning environment, especially when this learning environment is based on modern technologies and digital interfaces. If learning experience is, in fact, affected by individual characteristics, then the teacher's utility is also affected because the teacher's job is to facilitate the learning process. Given that learning preferences cannot be completely revealed to the teacher, then ways of improving student experience must be investigated.

Academic achievement can be the result of several factors, including prior academic achievement, motivation, subject-specific skills and preferences or styles in the ways students are inclined to approach a learning situation. It is indeed challenging but also significant to understand how individuals learn. A learning style or preference is usually considered as stable over time, but not always.

There are many types of preferences, which can affect the learning experience of a student. Types of preferences include: how the instructor approaches the delivery, the level of social interaction within a particular learning session, how an individual processes information, how the cognitive personality style affects learning, etc. The last one is only apparent if a student's behaviour is observed across many different learning situations, so it becomes very challenging for a teacher to incorporate a corresponding strategy in the lesson plan. However, a teacher can incorporate different modes of instruction, to address instructional preferences, or different approaches on how to explain new knowledge, to address different information processing styles.

Different preferences towards instructional approaches can be easily satisfied during the lesson. For example, a student's preference to listening to the lesson, and to take notes, or to have visual representations of the knowledge, such as charts or graphs, are some easy lesson plan approaches. On the contrary, a preference towards

© Springer Nature Switzerland AG 2021
J. Antoniou, *Quality of Experience and Learning in Information Systems*,
EAI/Springer Innovations in Communication and Computing,
https://doi.org/10.1007/978-3-030-52559-0_3

tactile processes is not always possible and even more challenging in digital class-rooms. When it comes to considering cognitive styles or preferences, i.e. how a student perceives and processes information, incorporating relevant approaches in the lesson plan is not so easy. Considering such information-perceiving and information-processing preferences can support a student's creativity, problem-solving, communication and decision-making skills.

In terms of cognitive styles, we will explore three cognitive styles, for which an indicator is available, referred to as CoSI (Cognitive Style Indicator), published in 2007 [3]. What we are interested in, for the analysis of the learners' experience, is the characteristics of the particular learning styles and not the indicator itself.

The first cognitive style that will be considered is referred to as the *knowing* style. The characteristics, preferred by the learner that processes information using this style, are rationality, use of logical, analytical and impersonal information. The second cognitive style is referred to as the *planning* style. The characteristics of this style are the use of planned organised routines, sequential and structured steps. Finally, the third cognitive style is the *creating* style, and as the name implies, the learner that processes information using this style prefers ideas, subjective approaches, open-ended questions, inventive and creative descriptions, an array of possibilities and meanings.

We need to make a note here of the popular VARK model, as another interpretation of learning styles. The VARK model was presented in a 1992 study [5], and it presents four types of learners after these were identified from extensive observation. The first type of learner, represented by the letter "*V*" of the acronym *VARK*, is the visual learner, i.e. a learner that better internalises and synthesises information when the information is presented in a visual manner such as a graph. "*A*" stands for auditory and characterises learners, who have a preference for information which is presented vocally, i.e. such that they can *hear* the information. Reading (and writing) learners are represented by the letter "*R*" and their preference is for written material, whether these are in-class handouts or text on the lesson's PowerPoints. They also better internalise and synthesise content when they have the opportunity to take notes or work on written assignments. Finally, "*K*" stands for kinaesthetic learners, and these are learners, who prefer to learn by "hands-on" activities, i.e. by participating physically in the learning process. This can be different from the tactile learners that were mentioned previously in the chapter, since the "touch" is not as an important aspect as the practical component. For such learners in a digital classroom, solutions of simulation environments can be very helpful for improved learning.

3.2   Learning Preferences and User Experience

There has been some research done within the field of education to explore the effect of such preferences on learning, and opinions vary throughout the years. Nevertheless, we do not aim to explore further the effect of such preferences on learning; instead, the chapter will focus on and investigate scenarios of how the learner's experience is affected. The chapter considers that when a learner has a preference that is not satisfied, the experience is degraded. In particular, we are concerned with the learner's quality of experience in a digital environment. When considering that the lesson takes place over a communication network and with the use of personal computing devices for all the students, particular details of the delivery of the lesson may need adjustment. For instance, a series of short videos could replace a longer face-to-face lecture, or a hands-on engineering experiment may need to be replaced with a simulation-environment-based experiment that can efficiently run over a public communications network. Nevertheless, even in these new learning environments, we consider learners to have preferences and that the quality of their experience can be affected by these preferences, regardless of the interface options or the characteristics of the communication network.

Moreover, a learner may prefer a combination of styles in terms of instructional mode, information processing styles, or types of interaction. This complicates understanding and assessment of these styles by the teacher. The teacher's role can be better characterised as the learning facilitator for the group of students. A teacher may assume that a variety of preferences exists within the group and should be flexible in terms of lesson preparation and lesson delivery. Although it is challenging to analyse such preference profiles for a group, we will try to approach the digital classroom from both a teacher and a student perspective and aim to characterise simple interactions and utilities, when such preferences are considered.

Considering such preferences can be very helpful in planning a lesson based on online learning, so that concepts can be presented in different ways and formats, so that the learning experience is enhanced. Online learning is often challenging because it lacks the immediacy that face-to-face classroom have, so the student experience should be supported in other ways. Learning styles and preferences have a place in such learning environments in order to better support the student audience. A carefully designed learning plan that considers such preferences may improve experience, in terms of motivating students to better engage with the lesson. Using this approach, individuals' experience is enhanced and hence, the element of immediacy that was lost by transforming the face-to-face lesson into an online one can be recovered. Thus, learning preferences are important to consider for a lesson's learning strategy, in terms of enhancing students' motivation to learn.

3.3   Learning Styles and e-Learning

We will briefly reference the importance of the quality of e-learning software and consequent digital pedagogies. Yiatrou et al. [20] recognised the complicated task of characterising learning and produced a set of heuristics that consider a number of learning theories. The pedagogical heuristics consider *theories related to technology and our current digital society,* and overall results of the study show that *e-learning can offer educational benefit.* The authors have attempted to use multimodal learning as part of their e-learning software, i.e. to represent the educational material through approaches that cater to various learning preferences, such as diagrams, videos, audio and textual representations, images as well as activities. This was a successful pedagogical approach, which resulted in positive comments from the students. The multimodal approach was juxtaposed to a single mode of content representation, which was not received favourably by all students. Overall, results of this study reinforce the opportunity to improve student experience through digital platforms and educational technologies, such that more of the students' preferences can be satisfied.

Digital pedagogical strategies need to be developed to support such online learning. A digital pedagogy is a set of tools for learning that explore the use of digital resources, their effectiveness and their impact. Such strategies also consider improving the teaching by the use of digital technology. Overall, digital strategies in education have the potential to make education more accessible and inclusive and support lifelong learning. We will approach the lifelong learning opportunities in a subsequent chapter.

A digital strategy is an opportunity for an educational institution to transform education, by planning for all students, making content and tools both available and accessible. Prior to considering the actual interfaces, it is important that the educational institution examines and improves its own digital infrastructure. Allow for trainings that are available to all students through a well-performing, efficient technology infrastructure. In addition, skilful personnel must be able to support the learners in using the digital infrastructure. Such personnel include administrators, technicians and educators. Another way to improve the experience of the learner is to allow for their feedback or contribution to the learning, so that the learners can actively experience the learning experience. Finally, revisit the design of the interfaces and ensure that the technology users are comfortable and satisfied with the quality of the interaction with the technology. We will return to this point in one of the following chapters.

We need to observe and understand new technologically enhanced learning environments and any tensions that the environments may create between teachers and learners. The purpose of the chapter is to consider the interaction between the teacher and the learners, given that they carry individual learning preferences, as well as to investigate tensions between learners themselves, in cases that they need to deal with aspects of the lesson that either align or not with their preferred learning styles.

3.4   Model Educational Scenarios

Consider a classroom, which is supported through a digital platform and attended by a group of students, who study through this distance-learning, online approach. Each student may have one or more learning styles or preferences, whether these align or not with a specific learning style model, such as the VARK model, or the Cognitive Style Indicator model. The preferences may in some cases pose special challenges for a lesson delivered through a digital classroom. We may consider that students have different preferences in terms of the actual preference but also the complexity of the combination of styles that they may prefer.

In exploring the interactions within the group attending the online course, both teacher and students, we make a set of preliminary assumptions:

- There is a finite number of students in a digital classroom.
- Quality of student experience is affected by students' learning preferences.
- Only preferences that may be satisfied in a digital classroom are considered, i.e. tactile learning preferences are not considered.
- Each student has a different set of learning preferences that vary in complexity.
- The teacher cannot in one single lesson address all learning preferences of the student cohort, and therefore some of the students will not be satisfied.
- The teacher wants to address all learning preferences of the students that participate in the digital classroom.
- There is a best and a worst pedagogical strategy approach that the teacher could employ to address the learning preferences of the student cohort.
- Since the aim is to improve experience for students and teachers, then there is a positive utility in addressing more learning preferences.
- Hence, the utility or *payoff* from successfully addressing more complex set of learning preferences is more when compared to addressing a less complex set of learning preferences.

Thus, we have the teacher that wants to address a set of learning preference sets, one per student, such that not only most of the students are satisfied but also the most complex sets of learning preferences are addressed.

Let us refer to the teacher's approach as a *mixed learning delivery strategy*. To better understand the teacher's challenging role, we will explore a model where we try to figure out what the best *mixed learning delivery strategy* is. We employ important assumptions for this model, such as the fact that the strategy is not expected to possibly address all preferences. Exploring this situation does not necessarily provide a recipe for the lesson plan but explores the teacher options and choices that can support a better outcome for the class in terms of learning and experience. Initially, we explore a model that generalises the situation and does not proceed to label the specific strategies. This helps us understand the teacher's perspective, i.e. given that the teacher does not know the specific sets of learning preferences that each student possesses.

This initial model of approaching the design of a *mixed learning delivery* strategy is followed by a second model, which explores the interaction of two learners with the lesson delivered, and with each other if there is opportunity for students to deliver lesson components. This is done in order for us to explore the student perspective. In the second model, we do present and investigate examples with specific learning preferences to better understand the student perspective, since the students are aware of their own preferences.

3.5   Using Game Theory for Scenario Modelling

We apply a game theoretic approach to modelling these assumptions and offer a representation of preferences and decisions. Given each perspective, i.e. the teacher and the students, we look for an equilibrium, i.e. a state from each perspective when all decision makers are happy and do not want to change strategy. A perspective is often referred to as a *player* in game theoretic models.

Game theory [6] is a mathematical tool that models decisions and strategies of interacting entities by considering actions and corresponding quantitative payoffs. It is therefore a great tool for modelling and analysing decision-making situations. However, game theory is not only based on mathematical elements to model a situation, e.g. by quantifying the payoffs. It needs to take into consideration the type of interaction, the options of the interactive entities and their expectation, as well as the logic behind the decisions, tapping into a bit of psychology or sociology in the case of group behaviour models, as well as a bit of philosophy and logic. As such, the field of game theory becomes an interdisciplinary approach to characterising decision-making situations.

Because of its interdisciplinary perspective, game theoretic tools have been used to model strategic situations in a variety of disciplines, including economics, political science, psychology, business and computer science. It offers a theoretical decision-making framework that can be used to represent and analyse situations where decisions or strategies need to take place. Game theory, otherwise referred to as *game of strategy* [4], appeared formally in the 1940s in a text by John von Neumann and Oskar Morgenstern [12], although the ideas of games and equilibria are found as early as 500 AD in the Babylonian Talmud, which is the compilation of ancient law and tradition for the Jewish Religion [2]. The 1950s and the 1960s was the time that game theory gained popularity because of the important contributions from John Nash [11], Thomas Schelling [17], Robert John Aumann [1] and John Harsanyi [8], followed by the 1970s important publications of Reinhard Selten [18]. All the aforementioned scientists were awarded Nobel Prizes for their work. More recent significant publications include, but are not limited to Ariel Rubinstein [15] who contributed in the theory of bargaining, Koutsoupias and Papadimitriou [10], for their work on equilibria and the Price of Anarchy.

One of the game theoretic examples, which will serve us in this chapter is the example, referred to as *Hagar's Battles* [6]. We will reproduce the example next and then proceed to draw parallels between the aforementioned example and our educational scenario.

For the *Hagar's Battles'* scenario, we consider that there are ten battlefields, with each battlefield carrying a different value for the fighting parties, such that all values are different. Therefore, there is one such battlefield with the highest value and one with the lowest vale. Each of the fighting parties has ten fighting units, e.g. ten soldiers. Each party can send at most one unit to each battlefield, and when a unit is sent to a battlefield, then it is assumed that the battlefield is occupied by the corresponding fighting party. The winner is the party whose fighting units have occupied a set of battlefields with higher cumulative value than the opponent.

According to the analysis of the specific example, game theorists have concluded that the specific interaction has a unique dominant strategy equilibrium. That means that each of the interacting parties will have a best set of moves that can offer the best possible payoff. In short, the analysis shows that when there is an opportunity to switch a unit from a battlefield with a lower value to a battlefield with a higher value, then there will always be an increase in the overall payoff with this switch. On the other hand, if there is not an opportunity to switch to a battlefield with a higher value, no switch will be made.

In a classroom setting, digital or otherwise, there is an opportunity for the teacher to design the lesson such that it can satisfy student preferences, in an attempt to enhance their experience. Given that one of the assumptions that we make for this example is that a single teacher cannot satisfy all students' preferences in one lesson, then we can draw parallels between different preference profiles and the battlefields themselves. We attempt to create a strategy, where the preferences that are eventually addressed by the teacher's lesson plan result in higher cumulative payoff than the ones that are not addressed, similarly to the *occupied* battlefields versus the battlefields that are *lost*. The following section offers the specific details.

3.6 Mixed Learning Delivery Strategy Scenario

Let us first proceed with the first example, where the design of the *mixed learning delivery strategy* is explored. Consider that even the best learning strategy cannot assume that it can address all students' preferences in a single lesson plan. We explore the modelling of the digital classroom scenario, from the perspective of the teacher trying to create the best *mixed learning delivery* strategy. The model will compare a strategy for a set of students that are satisfied and compare this with the corresponding set of students that will not be satisfied (as per the model's initial assumptions). Therefore, the utility of the students whose preferences are satisfied should always outperform the corresponding utility, which results from a strategy that does not satisfy the students' preferences.

Assume that there is a finite number n of students in the digital classroom such that each student has a different set of learning styles or preferences, starting from the simplest set of preferences for the first student, s_1 and moving up to the most complex set of preferences for the nth student s_n. Accordingly, a strategy S_i is more successful if it addresses a more complex set of learning styles than a less complex set of learning styles, such that the payoff values for this strategy $P(S_i)$ are:

$$P(S_i) = v(s_1) < v(s_2) < \cdots < v(s_{n-1}) < v(s_n)$$

Therefore, a strategy S_i, where i is the strategy's identification number, is a set of numbers selected from the set of whole numbers between 1 and n, such that the selected numbers represent the students that the mixed learning delivery strategy has satisfied in terms of learning styles or preferences. Hence, for the same game model, a strategy S_j is the set of all numbers that are not selected by S_i that represent the students that the mixed learning delivery strategy has not satisfied, for the specific digital classroom.

If we model a competition between the S_i and S_j strategies, then the complete sets of students participating in the classroom would be divided between the strategy solution sets for each strategy, i.e. some of the students would be satisfied by the mixed learning delivery approach and some would not be satisfied, as their preferences would not be addressed.

To be more specific, we need to assume that one teacher in one lesson cannot address all students' learning preferences. This will most likely be the case in most digital classrooms, especially with a large number of student audiences. So, in a competing model where the *best* strategy, S_i, will address some of the students' learning styles or preferences and strategy S_j will not address some of the students' learning styles or preferences, we can show that there is a unique dominant strategy equilibrium that the teacher may consider for the lesson plan. A dominant strategy is a strategy the *player*, in this case the teacher, will adopt no matter what the students' actions are given that the specific strategy will always dominate the alternative strategy.

Therefore, we want to show that a teacher will always try to satisfy most of the students' preferences in the best possible way, recognising the complexity of some of these preferences, and understanding that a complete satisfaction of the student audience is not always possible.

For the teacher to be satisfied from this game, the mixed learning delivery approach needs to be able to address students' preferences such that the sum of the values of students' preference sets satisfied is greater than the sum of the values of students' preference sets not satisfied.

Therefore, $P(S_i) > P(S_j)$, specifically:

$$\sum_{n}^{k=1} v\left(s_k\right), \forall s_k \in S_i > \sum_{n}^{m=1} v\left(s_m\right), \forall s_m \in S_j$$

We begin designing this model by a less optimistic approach, by assuming that the *best* mixed learning delivery strategy does not manage to address the most complex set of learning preferences for student n. Let $v(s_n)$ be the highest value set of learning preferences that has not been addressed by the best strategy. As a starting point, this is acceptable, as we assume that the teacher wants to address the most complex sets of preferences because the more complex the set, the highest the overall payoff value for this model. However, as we mentioned previously, we cannot assume that all sets will be addressed, so we define the most complex set as not addressed for this model.

Thus, s_n will end up as part of the set of students in S_j, and $v(s_n)$ will end up contributing to $P(S_j)$.

We may also assume that there will be one or more students and corresponding sets of learning preferences that the specific mixed learning delivery approach S_i can satisfy. Out of these, one will contribute the lowest to the overall payoff. The lowest value, or a lowest complexity set of learning preferences that the strategy can satisfy, can be the lowest value in the entire set of students, i.e. student s_1.

The teacher will always replace an approach to satisfy a less complex set of strategies with an approach to cover a more complex set of strategies. Given, the failure to satisfy the most complex set of strategies, the teacher approach should be to use any resources to actually satisfy that and give up effort or resources to satisfy the least complex of the preference sets.

Questions of how to recognise preference sets and revise the mixed delivery learning strategy cannot be answered without interaction with the specific student audiences. Some researchers have even dismissed the effect of learning styles on improving learning [14], but we are concerned with the effect of the student experience as a participant in the digital classroom, and hence an evaluator of the medium used to deliver the lesson, using the abundant capabilities of emerging technologies to enhance the field of education. Several researchers have approached such evaluation by collecting user opinion data and evaluating the digital interface of the online class for UX evaluation [13]. Instead, we will try to model and analyse scenarios and predict behaviour and decision-making as a game-theoretic exercise.

Let's assume that the teacher will update the mixed learning delivery approach to not address the set of learning preferences with the least complexity in order to allow resources for addressing the set of preferences with the highest complexity, which was not addressed previously. Therefore, the gain in the overall payoff with this revision is greater than 0, i.e. the decision to update resulted in a positive payoff, because the number of students satisfied will remain the same but the complexity of preferences addressed will increase, increasing the overall payoff for the strategy S_i.

Thus, as with the initial assumption, a teacher's dominant strategy is to always aim for the best delivery approach possible by focusing the effort and resources on addressing a more complex set of learning preferences. This approach will always result in a positive payoff and does not depend on any other actions or circumstances that may affect the *game*. From the perspective of the teacher, the payoff is the sum of all the student values that have sets of preferences that are addressed. Given that the teacher does not immediately design the perfect strategy, i.e. a strategy that addresses most or all sets of preferences, including the most complex set, then there will always be at least one change: a swap between a less complex and a more complex set of learning styles to increase the overall gain by a positive value.

It is important to note that there will be students that will not be able to gain as much as other students since the best strategy S_i will most likely not address all students' learning styles or preferences. Next, we attempt to examine this situation from the students' perspective.

3.7   Information Processing Styles Scenario

Let us assume that student A and student B attend the same digital classes. Their learning preferences are quite similar. Specifically, they enjoy the distance learning aspect of the classes, the auditory and visual components of the classes, especially because they are not fans of taking notes that would otherwise be necessary in a face-to-face class. They are happy that they can record their online sessions and watch them later for further studying the concepts.

However, their information processing styles are different. For the purposes of this model, we will consider the information processing styles described in the publication by Cools and den Broeck from 2017, entitled *Development and Validation of the Cognitive Style Indicator* [3]. There are three different types of information processing styles described: the *knowing* style, the *planning* style and the *creating* style. On the one hand, student A has a more of a *knowing* information processing style, a style characterised by the student enjoying more *analytical*, *logical* and more *objective* examples during lesson delivery. On the other hand, student B has more of a *creating* information processing style and prefers to learn by exploring *open-ended* examples, investigating *meanings* and *ideas* and enjoys examples during lesson delivery that allow more of a *subjective* perspective.

Both students find the examples during lesson delivery to be an important part of their learning. For a model of the students' perspective, we will consider that for a specific digital course, there is only time for one example per lesson. That limits the delivery approach such that it can only address either student A or student B learning preferences with reference to the delivery of the example, since it can either be delivered as a *knowing*-style aligned example or as a *creating*-style aligned example. Of course, we must consider that the teacher also has a third option with regard to the style that the example is delivered in. This is the third information processing style, the *planning* style, which both of the students partially enjoy, but do not prefer. Thus, the example can also be delivered in a *planning*-style aligned manner.

Since the example is an important part of the learning in a particular lesson, both students need to try to understand it. Understanding and enjoying the example contributes to their payoff. Therefore, on the flip side, the most effort the students need to put into understanding it, the less the satisfaction.

Furthermore, we may stipulate that in case the example is not delivered in the student's preferred information processing style, the learning itself is suboptimal. The overall student payoff increases as: (i) the learning increases and (ii) the effort to understand it decreases. Let the maximum learning L be achieved if the example is given in the student's preferred style, and a suboptimal amount of learning L' be achieved otherwise. Thus, the overall payoff for each student is affected by the students' alignment with the employed learning style delivery approach.

If the learning style delivery method employs a *knowing*-style alignment, student A will receive payoff $P_A = L - e$, where e is the effort needed by student A to follow the example, which should be minimum but non-negligible. On the other hand, if the learning style delivery method employs a *knowing*-style alignment, student B would receive minimal learning satisfaction as it would take an amount a of effort to understand the example, where $a > e$. Moreover, given that the example is not aligned with the student's preferred style, the learning would be $L' < L$. Therefore, for student B, $P_B = L' - a < L - e$. Therefore, for a delivery aligned to a *knowing* information processing style:

$$P_A > P_B = L - e > L' - a$$

If the learning style delivery method employs a *creating*-style alignment, student B will receive $P_B = L - f$, where f is the effort needed by student B to follow the example. Given that *creating*-style examples are often open-ended and require student participation, we consider that in relative terms $f > e$, but overall f is minimal but non-negligible for student B. On the other hand, if the learning style delivery method employs a *creating*-style alignment, then it is not aligned with student A's preferred information processing style. Thus, the learning achieved by student A would be $L' < L$. Furthermore, student A would require an amount a of effort to understand the example, since the two styles preferred by the two students are considered opposites and it would take a great effort to understand an example delivered in each other's preferred style. We may consider this effort to be equal in both directions. The payoff for student A would be in this case $P_A = L' - a < L - f$. Therefore, for a delivery aligned to a *creating* information processing style:

$$P_B > P_A = L - f > L' - a$$

Finally, if the learning style delivery method employs a *planning*-style alignment, both student A and student B will need to make an effort g such that $g > f > e$ in order for them to understand the example. Hence, $P_A = P_B = L' - g$, for both student A and student B and $L' - g < L - e$ for student A and $L' - g < L - f$ for student B. Let the *planning* style for information processing be closer, in terms of relevance and familiarity, to the other two information styles, i.e. *knowing* and *creating*, than they are to each other. Thus, the effort a needed to understand each student's preferred style by the other student is eventually more than the effort g needed by both students to understand the *planning* information processing style, such that $L' - a < L' - g$.

Thus, each student would be better off with an example delivered in a *planning* information processing style, rather than with an example aligned with the other student's preferred learning style. In case the example is given in the opposing information processing style, then each student can try to learn by translating the example into a *planning* style of an example in order to better understand it. This is an alternative action to trying to understand the example in its current form, be that *knowing*-style aligned or *creating*-style aligned, given that an opposing style is the hardest to understand by both students. In that case, the learning would still be suboptimal, i.e. L', but the effort would be g' such that $g' > g$ (to account for the translating effort), but still less than effort to understand an example in an opposing style previously labelled as a. Therefore, student A may apply a *planning* approach to understand a *creating*-style aligned example and student B may apply a *planning* approach to understand a *knowing*-style aligned example. Eventually, both students would end up with a payoff $P_A = P_B = L' - g'$.

The students may also try to translate the *planning*-style aligned example into their preferred style by also using an effort comparable to g'; however, the learning would be L in this case since translating in a student's preferred information processing style provides the maximum learning. Concluding, we may also note that it is not expected that either of the students will try to translate an example into the opposing information processing style, as this would result in the least amount of payoff. This is indicated by zero (0) in the payoff matrix below. Note that all other payoffs are greater than zero (>0).

A tabular representation of this model's payoff matrix is found below (Table 3.1).

There is a clearly dominating strategy for each student if the example is given in one of their preferred styles. However, the payoff is suboptimal in the cases that the example is presented in any other style. The student perspective demonstrates the variation of decisions that a student has and the corresponding variation of payoff (in terms of student quality of experience; note that we do not attempt to quantify learning but the subjective satisfaction from the learning experience). Another conclusion that is evident from this simple example, i.e. simple in terms of the range of styles considered, is that there exists complexity and challenge in putting together a lesson plan.

3.8   Flipped Classroom Scenario

One of the goals of the use of technology to deliver education is to leverage the technological infrastructure by making it accessible to teachers and students in order to allow new ways of delivering learning, and exchanging resources and ideas with the use of practical digital tools developed to support this new teaching and learning mode. The consideration of digital means for learning does not only promote digital innovation but may also allow for an easier adoption of alternative pedagogical methods, aiming to support even more learning styles and preferences and eventually enhance student learning. In fact, given the immediacy of the use of this media, a great opportunity emerges for the teachers to re-evaluate and re-position the primary role, moving away from simply a teaching role and towards a

Table 3.1 Payoff matrix for students with different information processing styles

Payoffs (Student A, Student B)				
Row labels: Information processing style example is presented in		Knowing approach	Planning approach	Creating approach
	Knowing style	$(L - e, 0)$	$(L' - g', L' - g')$	$(0, L - g')$
	Planning style	$(L - g', L' - g)$	$(L' - g, L' - g)$	$(L' - g', L - g')$
	Creating style	$(L' - a, 0)$	$(L' - g', L' - g')$	$(0, L - f)$

role of facilitators of learning. With this new role, the consideration of learning styles may receive even more importance than in traditional delivery modes, i.e. in classroom, face-to-face education.

In the example above, we modelled the perspective of the student by considering the delivery of the lesson by the teacher, but the new media may offer more opportunities to employ pedagogical methods where the student takes control of the lesson delivery, or at least of lesson components. This allows the simplification of redesigning learning by the teachers to promote different perspectives and learning by teaching. The reason that this becomes simpler is the familiarity with the digital interface by all participants, the ease with which control may be shared between students and teachers in such environment (e.g. screen sharing) and the apparent decrease in the distance between the student and teacher roles during lesson delivery. As a result, the teachers' relation to curricula changes and similarly the students can feel more in control of their learning.

By sharing control with the students, the teachers can use the platform to further facilitate peer-to peer-learning, e.g. through a peer to peer mentoring scheme, or by adapting the *flipped classroom* pedagogical model, and allowing students to be in charge of the teaching of particular lesson components, to enhance their learning. Such components may include practical examples or scenarios that demonstrate use of theoretical concepts. Given the model of student A and student B learning through a particular practical example, we will use this in the following discussion as well.

Before going into the model itself, it is useful to offer some discussion on the *flipped classroom* concept and on how this can be useful in a digital classroom. The interested reader can further explore this through a number of content specific examples in a book by Abigail G. Scheg, entitled *Implementation and Critical Assessment of the Flipped Classroom Experience* [16]. The *flipped classroom* model focuses on learner-centred delivery. Given that digital learning environments allow multitasking and that students themselves may be more familiar now with such technologies than past students' generations, it is expected that lesson delivery can take advantage of this enhanced but familiar technological interfaces to improve student learning. In fact, a *flipped classroom* offers new opportunities to the students to engage with the class. The students can present practical components of the class, such as examples and exercises, that may have usually been part of the homework. This can take place under teacher supervision, but the traditional lecture-style delivery that was traditionally done in class is now done at home in preparation for the more active learning session of the *flipped classroom* [19].

The consideration of diverse learners, i.e. learners with different cognitive styles and different learning preferences is one of the foundations of a *flipped classroom*, and so is peer-to-peer learning, i.e. students taking over the lesson instruction [7, 9]. Of course, it is important to acknowledge some of the initial challenges of the digital model of the *flipped classroom* as there have been concerns that this will widen the digital divide, especially for students that are not familiar with such technology, which may put them at a disadvantage against their classmates or other students that are more familiar. However, as more and more courses offer digital components and technology increasingly supports more and more schools to provide digital options to their students, we expect that such concerns will not be significant in the near future.

Going back to our example of student A and student B, in a model of a *flipped classroom* either student A or student B would have the opportunity to present the example to the rest of their classmates as the instructor for the particular lesson component. We may assume that both students would prefer to present examples in their own preferred information processing styles or at least in a style they somewhat enjoy rather than use an information processing style, which they find most challenging.

The payoff from this activity needs to consider the effort of preparing the delivery by the student responsible for it. If the student delivering the example is student A, then the cost of preparing the lesson is e because student A plans to prepare a logical example with objective results that will not consist of any open-ended questions and discussion of ideas. If the student delivering the example is student B, then the cost of preparing the lesson is f, where $f > e$ because student B plans to prepare an example that will engage the student's classmates. Specifically the example will consist of open-ended questions and discussion of ideas and meanings, requiring clearer instructions and a plan for collecting the classmates' feedback and introducing it in the example. Both students' preparation efforts, e and f, are minimal for the students, since they are working according to their preferred information processing style, but they are non-negligible.

Both students will receive learning L, if they are the ones teaching the example, because preparing to teach it offers them maximum learning opportunity. Thus, as instructors, student A and student B will receive the following payoffs:

$$P_A = L - e$$

$$P_B = L - f$$

The two students may also decide to present their example, using a different style than their preferred style to achieve better collective learning, As mentioned previously, the *planning* style is categorised as an information processing style that is more familiar for students with preference in either *knowing* or *creating* styles, than those styles are to each other. So, by employing the *planning* style approach for the example, more students would easily be able to familiarise with it and better collective learning would be achieved. Let us call this approach a second option for both student A and student B. The payoff to both students would consist of the learning achieved minus the preparation effort. As instructors, the students would design the example in a way that they fully understand it before structuring it in a way to present it according to the *planning* style approach. Therefore, we consider that maximum learning is achieved but a translation to the specific style needs to take place in addition to the preparation effort. The payoff for both students would therefore be:

$$P_A = P_B = L - g'$$

(g' costs a bit more than just understanding a *planning* mode example with an effort of g, since translation into this mode is necessary; $g' > g > f > e$).

Now, given which of the two students becomes the instructor, we need to consider the payoffs of the student that will be the recipient of the lesson.

Let student A be the instructor. Then, student A may decide to provide the example in *knowing*-style aligned delivery or in *planning*-style aligned delivery. Accordingly, student B will have the following responses:

(a) In response to a *knowing*-style aligned delivery, student B would receive minimal learning satisfaction as it would take an amount a of effort to understand the example, where $a > e$, i.e. the effort to understand by student B is greater than the effort by student A to prepare it. Moreover, given that the example is not aligned with the student's preferred style, the learning would be $L' < L$. Therefore, for student B, $P_B = L' - a < L - e$. As a secondary action, student B can try to learn by translating the example into a *planning* style aligned example, in order to better understand it. This is an alternative action to trying to understand the example in its current form, given that an opposing style is the hardest to understand by both students. In that case the learning would still be sub-optimal, i.e. L', but the effort would be g' such that $g' > g$ (to account for the translating effort), but still less than the effort needed to understand an example in an opposing style previously labelled as a. There can also be a translation into a *creating*-style aligned example with the same effort but better learning. The payoffs for translating are:

  (i). Into a *planning* aligned example: $P_B = L' - g'$
  (ii). Into a *knowing* aligned example: $P_B = L - g'$

(b) In response to a *planning*-style aligned delivery, student B will need to make an effort g such that $g > f > e$ in order for the student to understand the example. Hence, $P_B = L' - g$, for student B and $L' - g < L - f$. for student B.

Let student B be the instructor. Then, student B may decide to provide the example in creating-style aligned delivery or in planning-style aligned delivery. Accordingly, student B will have the following responses:

(a) In response to a *creating*-style aligned delivery, the learning achieved by student A would be $L' < L$. Furthermore, student A would require an amount a of effort to understand the example. We may consider this effort to be equal to the effort required by student B to understand the example if presented in a *knowing*-style aligned manner. The payoff for student A would be in this case $P_A = L' - a < L - f$. As a secondary action, student A can try to learn by translating the example into a *planning* style aligned example in order to better understand it. This is an alternative action to trying to understand the example in its current form, given that an opposing style is the hardest to understand by both students. In that case, the learning would still be suboptimal, i.e. L', but the effort would be g' such that $g' > g$ (to account for the translating effort), but still less than the effort needed to understand an example in an opposing style previously labelled as a. There can also be a translation into a *knowing*-style aligned example with the same effort but better learning. The payoffs for translating are:

   (i). Into a *planning* aligned example: $P_A = L' - g'$
   (ii). Into a *knowing* aligned example: $P_A = L - g'$

(b) In response to a *planning*-style aligned delivery, similarly to student B described above, student A will need to make an effort g such that $g > f > e$ in order for the student to understand the example. Hence, $P_A = L' - g$, for student A and $L' - g < L - e$.

As discussed previously, the *planning* style for information processing may be considered to be closer, in terms of relevance and familiarity, to the other two information styles, i.e. *knowing* and *creating*, than they are to each other. Thus, the effort a needed to understand each student's preferred style by the other student is eventually more than the effort g needed by both students to understand the *planning* information processing style, such that $L' - a < L' - g$. Thus, each student would be better off with an example delivered in a *planning* information processing style, rather than with an example aligned with the other student's preferred learning style.

In case the example is given in the opposing information processing style, then each student may also translate the example into a *planning*-style aligned example in order to better understand it. Secondary actions described above allow student A to apply a *planning* approach to understand a *creating*-style aligned example and student B may apply a *planning* approach to understand a *knowing*-style aligned example. Eventually, both students would end up with a payoff $P_A = P_B = L' - g'$.

If the students try to translate the *planning*-style aligned example into their preferred style by also using an effort comparable to g', they would improve the learning, which would be equal to L in this case since translating in a student's preferred information processing style provides the maximum learning. It is not expected that either of the students will try to translate an example into the opposing information processing style, as this would result in the least amount of payoff.

3.9   Discussion

This chapter has discussed the user experience for learners interacting in a digital classroom by approaching the experience using three different model scenarios, including designing a lesson delivery strategy for the class, considering cognitive styles from the student perspective in learning and approaching the learning environment from a flipped classroom perspective.

The chapter tried to analyse the ways in which individual student characteristics and preferences may affect the student's experience by simplifying or complicating the learning experience. Digital environment must support the learning experience to provide a sense of comfort to the students. Such e-learning environments are based on modern technologies and digital interfaces, and are designed according to established pedagogical strategies.

Overall, digital strategies in education have the potential to make education more accessible and inclusive and support lifelong learning. We will approach the lifelong learning opportunities in a subsequent chapter. The scenarios show that it is possible to achieve an enhanced user experience in such environments even in situations that may employ complexities in the student cohort's preferences, both for the learning facilitator and the learners themselves, whether they are simply receiving the knowledge or in the case of a flipped classroom, actively learn by contributing to the lesson.

References

1. Aumann JR (1960) Acceptable points in games of perfect information. Pac J Math 10: 381–417
2. Aumann JR (2003) Game theory in the Talmud. Jewish law and economics research bulletin series. Bar-Ilan University, Ramat-Gan
3. Cools E, Van den Broeck H (2007) Development and validation of the cognitive style indicator. J Psychol 141(4):359–387. https://doi.org/10.3200/JRLP.141.4.359-388
4. Dixit A, Skeath S (1999) Games of strategy. W.W. Norton & Company, New York
5. Fleming ND, Mills C (1992) Not another inventory, rather a catalyst for reflection. To Improve the Academy 11:137–155
6. Gintis H (2000) Game theory evolving: a problem-centred introduction to modelling strategic behaviour. Princeton University Press, Princeton, ISBN 0-691-00942-2
7. Hamdan N, McKnight P, McKnight K, Arfstrom KM (2013) A review of flipped learning. Flipped Learning Network, George Mason University, Virginia
8. Harsanyi JC (1967) Games with incomplete information played by Bayesian players. Behav Sci 14:159–182
9. Johnson G (2013) Student perceptions of the flipped classroom. Thesis for Master of Arts in Educational Technology, The University of British Columbia, The College of Graduate Studies
10. Koutsoupias E, Papadimitriou C (2009) Worst-case equilibria. Comput Sci Rev 3(2):65–69
11. Nash JF (1950) The bargaining problem. Econometrica 18(2):155–162
12. von Neumann J, Morgenstern O (1944) Theory of games and economic bahavior. Princeton University Press, Princeton
13. Reid R et al (2016) Asking students what they think, student user experience (UX) research studies to inform online course design. In: Proceeding of Association for the Advancement of Computing Education (AACE): e-Learn 2016
14. Riener C, Willingham D (2010) The myth of learning styles. Change 42(5):32–35
15. Rubinstein A (1982) Perfect equilibrium in a bargaining model. Econometrica 98(1): 97–109
16. Scheg A (2015) Implementation and critical assessment of the flipped classroom experience, A volume in the Advances in educational technologies and instructional design book series. IGI Global, Hershey, 318 pages
17. Schelling TC (1960) The strategy of conflict. Harvard University Press, Cambridge, MA
18. Selten R (1975) Reexamination of the perfectness concept for equilibrium points in extensive games. Int J Game Theory 4:25–55

19. Sohrabi B, Iraj H (2016) Implementing flipped classroom using digital media: a comparison of two demographically different group perceptions. Comput Hum Behav 60(2016): 514–524
20. Yiatrou P et al (2016) The synthesis of a unified pedagogy for the design and evaluation of e-learning software for high-school computing. In: Ganzha M, Maciaszek L, Paprzycki M (eds) Proceedings of the 2016 federated conference on computer science and information systems, ACSIS, vol 8. Polskie Towarzystwo Informatyczne/IEEE, Warsaw/Los Alamitos, pp 927–931

Chapter 4
Cloud Computing: Considering Trust as Part of the User Quality of Experience

4.1 Introduction

The chapter explores the interaction between a cloud provider and a cloud user from the perspective of trust. Trust between the two interacting parties can affect the user quality of experience (QoE), especially over a recurring interaction, or, an interaction that is continuous over time. We are mostly concerned with the perspective of the user in this chapter, and this is the reason that the chapter will tend to focus on the user utility mostly. The cloud provider is also significant in our discussion. As the controller of the cloud service provision, the provider plays a role in inspiring trust and improving the user's QoE.

The chapter will primarily look at the cloud technology and some of its important characteristics in order to explain the reasons behind the question of trust and why cloud users may be sceptical about whether a trustful interaction with the cloud providers can be achieved. Then, the chapter will try to break down this interaction through a numerical model, designed as a game theoretic model, to show whether such trustful interaction can easily be achieved theoretically. The chapter does not provide extensive technical detail for the cloud technology, but publications are referenced, in order to direct the interested reader to sources of further reading. Technical concepts and meanings are explained where this is relevant to the interaction model and in order to provide a description of the environment details where the interaction takes place.

© Springer Nature Switzerland AG 2021

J. Antoniou, *Quality of Experience and Learning in Information Systems*,
EAI/Springer Innovations in Communication and Computing,
https://doi.org/10.1007/978-3-030-52559-0_4

4.2   Cloud Services

Cloud platforms are becoming more and more popular as they offer several advantages, especially to business owners. Advantages include lower costs for hardware, elasticity of resources and the ability to plug into many available cloud services. This seemingly should improve user experience, as the user may experience more and better quality services, due to the availability of resources. However, there are still some risks to consider that may affect user quality of experience. For instance, by using the cloud, the users give up control of their data, which is stored in an unknown location, and possibly on the same machine with other users' data. Whether this is something to consider as a threat to data security, is a concern of many potential cloud users.

Storing data, streaming video, or even hosting a website are services that usually require managing both hardware and software, especially at a business rather than at an individual level. This can be made easier with cloud computing, simplifying the management of both hardware and software for the cloud users. Cloud computing is about renting resources, such as storage space for example, or even networking resources and compute power. Most often, in addition to the above-mentioned types of resources, cloud platforms offer access to attractive service capabilities such as data analytics.

Storage service itself may include different types of services, for example access to storage space for unstructured files or access to storage space to create databases. With regards to networking service, access to networking resources may include secure connections with the cloud platform provider. When referring to compute power through the various types of compute services, this may include deployment of servers or applications. Another resource is often an analytics service. Analytics usually offer options for collecting and visualising performance data. The management of all these resources is simplified for the cloud user because the cloud platform provider is responsible for the physical hardware, and sometimes the software as well, keeping both hardware and software up to date.

With regards to business owners, the goal of cloud computing is to make running a business easier and more efficient. Since each business is unique and may have different needs, cloud platform providers offer a wide range of compute services to support the move of the business infrastructure to the cloud. The use of cloud platform for business has been an interesting topic recently. Some proposed articles for the interested reader include: "*A categorisation of cloud computing business models*" [3] and "*The development that leads to the cloud computing business framework*" [4].

When building solutions using cloud computing, a business owner can choose how the move to the cloud should be done based on the business resources and needs. For example, if a business owner wants to have more control and responsibility over maintenance, then a number of *virtual* machines should be created to host the business servers, and other business support systems, on the cloud. A virtual machine is an emulation of a computer device just like a desktop or a laptop computer. Each virtual machine includes an operating system and hardware capabilities that appear to the user like any other physical computer. Once the virtual machine is created, any other required software can be installed, and it will run on the cloud. This is known as *Infrastructure-as-a-Service* (IaaS).

Now, if the business owner does not want to use virtual machines (less maintenance and less responsibility), then containers can be used, where the cloud is used as a platform, where development can be done without needing to set-up the machine from scratch. Containers are similar to virtual machines except that they don't require a guest operating system and they provide an isolated execution environment for applications to run. This is known as *Platform-as-a-Service* (PaaS).

Similarly, serverless computing can be used, where development can be done without setting up the server software. Serverless computing can even simplify the execution per function of an application as this is triggered by some event. This is known as *Function-as-a-Service* (FaaS). The fourth type of generic service a cloud provider offers is *Software-as-a-Service* (SaaS), where applications can be directly used without configuration set-up or development requirements.

Figure 4.1 shows what the user can control for the different types of services, i.e. IaaS, PaaS, FaaS and SaaS. Each type of service has a trade-off; the more items the user controls, the less the management support received from the cloud provider. The user must decide on the type of service to use, based on the willingness to monitor, manage and maintain the corresponding cloud service components. As we can see from Fig. 4.1, some components allow both user and the provider to partially control configuration of these components.

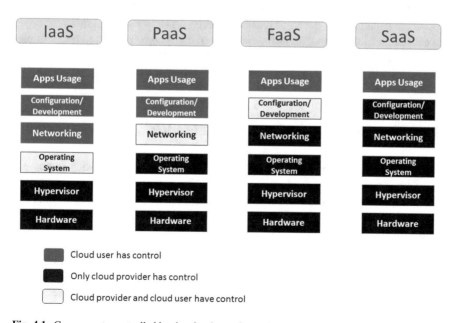

Fig. 4.1 Components controlled by the cloud user for each service type

Specifically, for IaaS, the user can create virtual machines from scratch, by selecting versions of operating systems made available by the cloud provider. Once that is configured, the user is responsible to set-up and manage the networking, i.e. configurations that control communication with the virtual machines. Also, the user can edit any configuration necessary for the virtual machine instance to run the required services, either by developing software or using the available cloud services and applications.

For PaaS, the user can configure the environment instance that would allow for the development of new software. The container environment allows for the user to control any client – server communication for the developed software – but does not allow access to the operating system. The operating system is controlled by the provider.

For FaaS, similar access control is given to the user, except the ability to edit any configuration related to the server side of the software development, because FaaS only allows for client-side implementation. Since only the client side software is developed, any client-server communication is unnecessary for the user. Thus, networking configuration is controlled by the provider.

Finally, for SaaS, the user only has access to specific application and cannot change any system configurations. The applications are created by third parties and made available over the Internet through access to the cloud platform. No additional configuration or update is necessary by the user.

In addition to compute services described above, storage services are very important for motivating the business owner to move to a cloud-based infrastructure. Most devices and applications need to read and write data. For each different type of application, the type of data, and, how it is stored can be different. Cloud platform providers usually offer different types of storage services such that different types of data can easily be handled. Different types of data can be addressed for example, by providing files, disk or databases. The advantage of using cloud-based storage is that it is usually elastic, i.e. it can scale to meet the business needs.

4.3   Cloud Pricing and Deployment Models

Usually cloud platform providers offer flexible pricing models [2], such as *pay-as-you-go* or *consumption-based* pricing models. This helps the business owner avoid any upfront infrastructure costs that would otherwise be incurred and pay only for the resources used at any time. The flexibility of the pricing models is a result of the ease with which resources can be added to or removed, either by the user or automatically (as per the user's preference) to the originally committed cloud resources.

Cloud platform providers usually offer the capability to scale resources horizontally and vertically. Horizontal scaling is the process of adding more servers or services to the existing ones and allowing them to function together as a unit. Vertical scaling is the process of adding resources to increase the power of one specific server or service. Elasticity is a very attractive feature for the business owner and oftentimes a necessary characteristic of cloud platforms. The workload that a business may have for its servers could increase or decrease based on user demand but having the server on the cloud can compensate for such variations. The cloud platform can automatically add or remove resources. Even the geographical location of the cloud servers can be adjusted as most cloud platform providers have data centres located in different geographical regions globally.

In terms of deployment models, there are three different cloud deployment models: the public cloud, the private cloud and the hybrid cloud. A cloud deployment model defines where the data is stored and how the cloud users interact with it. Public, private hybrid and community deployment models exist. A decision of which deployment model to use depends on how much of your own infrastructure you want or need to manage [7].

The *public cloud* is a type of cloud hosting that allows the accessibility of systems and of its services to the cloud users easily, e.g. using the Internet through a web browser. This type of cloud offers service models that are open for use.

The *private cloud* is a type of cloud that allows the accessibility of systems and services within a specific boundary, e.g. of the organisation that deploys it. For a private cloud, the platform is implemented in a secure environment monitored by the particular organisation.

The *hybrid cloud* is a type of cloud, which is integrated, i.e. it can be a combination of private and public cloud servers that are combined as one architecture but serve individual purposes. Non-critical tasks such as development and testing can be done using public cloud, whereas more critical tasks such as organisation data manipulation can be done using a private cloud.

One last type of cloud is the community cloud model. The set-up of a community cloud is shared among different organisations that belong to the same community, such that they share similar computing concerns.

4.4 Cloud Security

Even though there are so many advantages of using the cloud, there are still questions about replacing physical resources with virtual ones. Virtualisation is preferred because it provides significant cost savings, but at the same time, it is very difficult for the user to have a complete understanding if how virtualisation works from a security point of view. On the one hand, it can be argued that virtualisation is more secure than traditional deployment environments because of the isolation between the individual virtual machines and the fact that there don't seem to be that many attacks on hypervisors [9]. On the other hand, virtual environments are based on physical deployments and therefore, virtual environments should require security similarly to a traditional deployment environment, i.e. a physical deployment.

However, given that the new environment is even more complex than a traditional environment, it is important to understand that new security approaches are needed in addition to the traditional ones. The complexity of the environment originates from the fact that the new virtual machines end up creating new networks in addition to the physical networks they inhabit. Thus, it may be argued that the process for virtualisation security must include at least the following: security control for each virtual machine, security control of the underlying physical infrastructure, proper isolation of virtual machines so that any attack does not affect more than one machine and security control of hypervisor software [8].

Security concerns can affect more than just the virtual machines [5]. Cloud platforms often inter-operate in collaboration with other data generating systems. One such example is the Internet of things (IoT), which is integrated with cloud platforms to be able to manipulate large amounts of generated data into offering a number of specialised services. Examples of these services include services and applications for smart environments.

The IoT allows for participating "things" or objects to be available over the Internet. These objects were not traditionally active online or accessible. Examples include home appliances and cars. Different types of communications are necessary in such systems to make this "web of things" a reality, such as machine-to-machine communication. In such communication models, data can travel through the network without any human intervention, i.e. generated, transmitted and received by objects or machines. Such automated communications can pose unique security risks to the network.

For instance, devices participating in the IoT can easily be infected with malicious software in order to be controlled by authorities that are not the object owners. The infected devices can in turn be used without the knowledge of the devices' owner(s) and forced to act as a set of transmitting devices on behalf of the hackers. One way to safeguard against such attackers is to have good device authentication mechanisms. It is often the case that owners of smart devices and IoT-connected appliances do not consider the significance of having strong security when connecting these devices to the Internet, even if it is as simple as making sure that a strong password is set. It is important to note that strong passwords must be set and updated as often as possible [1].

4.5   Trust in Cloud Interactions

Oftentimes, trust and security go hand in hand, such that the concerns for secure cloud interactions may result in lack of trust for the cloud platform providers. The idea of *hiding* the infrastructure leaves room for doubt. As with any service provider, a cloud service provider will make *promises* through agreements with the user, and consequently, by agreeing to the service contract or *promise,* the cloud user offers to *trust* the cloud provider.

Overall, the interaction between cloud providers and cloud users is achieved through service agreements, where the cloud provider offers to handle some of the security risks and the cloud user may decide to agree or not [6]. Cloud resources and cloud technologies are offered to potential users as a service. As the *customers* of this service, the cloud users expect that their interaction with the cloud service will be secure, e.g. that any data they own and is stored on the cloud will not be accessed by the cloud provider or a third party, etc.

The expectation comes from the need to sign service agreements with the cloud provider to ensure that any private data and activity will be safeguarded. The service agreements offer more data, but we will examine the information in such agreements that pertains to security specifications, as this is one of the main areas that the cloud user may experience lack of trust when interacting with a service provider. The reason is that cloud users will not be able to monitor the physical infrastructure themselves or traffic to the cloud in general, an agreement or contract is needed to guarantee security.

The agreement between the cloud provider and the user may include terms that reflect a detailed set of security measures that are indeed taken and can allow the cloud infrastructure to recover from possible malicious attacks against the cloud or the user or any system failures because of any other non-malicious causes. For example, some mechanisms that could be put in place are:

- Secure frequent backups to allow data retrieval if necessary.
- Intrusion detection mechanisms to recognise and avoid malicious traffic.
- Enforcing the user to authenticate correctly and frequently all of their cloud resources.
- Enforcing encryption of communication channels.
- Minimising privileged access.
- Securing APIs.
- Monitoring cloud activity for data loss.
- Offering training opportunities for the users.
- Etc.

Some of these measures are useful but non-compulsory for the cloud provider according to regulation. Therefore, the cloud provider can outline security measures in less detail, allowing for some risk but ensuring some cost-savings relevant to implementation. The measures that could be avoided may have effect on individual users but not the cloud overall infrastructure and service. Without compromising overall security, the provider can offer a service agreement that describes a seemingly risk-free set-up in all aspects, or a secure but more risky set-up, where often the user is expected to implement some of the security such as protection from phishing attempts or password cracking attacks.

Eventually it comes down to the user accepting the agreement, which can also be viewed as a *contract* between the cloud provider and the cloud user, as either *risk-free* (or acceptably risk-free) or *risky*. The interaction can be modelled as a game theoretic model, which we will call the *trust game model*, given that a cloud user would need to decide whether or not this contract can be trusted. By referring to the model as a game theoretic model, we refer to the attempt to characterise the utilities of two interacting entities such that a solution can be reached where both entities will gain as much as possible from the interaction.

The first entity is the cloud user, who relies on the contract itself as there is no way of monitoring the actual traffic to verify or validate the actions that the cloud provider commits to. The trust game model has an element of chance in it; there is a probability that the contract will be valid and complete, i.e. that the cloud provider takes all necessary measure, and a probability that it is not. Moreover, there is a probability that something will indeed happen to reveal the incomplete nature of a risky agreement.

The model does not imply malicious act but consider that there exists a set of actions that can be taken by the cloud provider to make the contract a truthfully risk-free contract (or the cloud experience truly secure). Legality and good professional practice are not in question; the only question is whether the user's perception or expectation of the contract is satisfied. Note that the term "true" onwards refers to the risk-free matching of the user's expectation, whereas "untrue" refers to a risky contract.

4.6   The Trust Game Model

4.6.1   Introducing Nature in the Model

Let us refer to the aforementioned probability of something happening to reveal a risky contract as *Nature*, such that Nature generates a probability of realised risk, P_t. This probability is external to the cloud provider and the cloud user, i.e. this may have to do with network vulnerabilities, or probability of physical disasters, the geographical region and physical cloud infrastructure is located, other financial conditions, legal frameworks across different locations of the world and many other factors. The probability is known to both the cloud provider and the cloud user, but the revelation occurs as the interaction progresses and not ahead of the point in time when decisions are made.

Given this generated probability, the user must decide whether to trust the interaction with the cloud provider and the agreement between them as *true,* or whether to discard it as *untrue*. Given the user's decision, then the user QoE will be affected *positively* if the user was correct in their decision, and *negatively* otherwise.

As discussed, the user cannot monitor cloud traffic; however, we may assume that it is possible to verify whether or not the trust in this contract was correct. The question of whether it is possible to verify or not this decision practically is not so significant as we may assume that in time this can be verified by the user's level of satisfaction during continuous interaction with the cloud service over a long period of time. We do not view this as a repeated interaction, which is a model we will explore in subsequent chapters. This is a normal form game, where the decisions carry a payoff that is revealed once *Nature* makes its play in the game, i.e. verifies the risk or not.

Therefore, the cloud user must make a decision that in essence predicts the long-term satisfaction from this interaction as it relates to the risks of the agreement or contract. The result of this decision is the payoff for the cloud user, i.e. the user is happy and receives a positive payoff if the decision made at the beginning of this interaction eventually matches the reality.

4.6.2   Understanding the User Perspective

Let the cloud user payoff be positive if the user is right and zero if the user is not, such that payoff $P_{user} = R > W$, where R is the positive payoff in case the user was right during initial interaction, and W equals to zero payoff and represents the user being wrong during initial interaction.

Accordingly, we expect that the user payoff is positive if the user was right, i.e. if the user evaluated the contract as risk-free, then no security violations were detected by the user, and if the user evaluated the contract as risky, then there was evidence of security violations that did not affect the user. Similarly, the user payoff is not positive if the user was wrong, i.e. if the user evaluated the contract as risk-free, but security violations were later detected by the user, and if the user evaluated the contract as risky, there were no later evidence that confirmed this.

We are focusing on the user perspective and we relate the user payoff to user QoE, such that the higher the payoff is the more satisfactory the user QoE is expected to be. Note that we are evaluating this from the perspective of the user, so the QoE does not depend on *Nature's* play but on the user's perception of how *Nature* played, or alternatively, the user's perception of whether the cloud provider has indeed provided a risk-free contract. We will explain this in more detail as we later discuss all possible decisions that the user makes.

Combining this with Nature's play and the actual contract contents, then in case the contract proves to truthfully be *risk-free*, the user receives R if the user evaluated the contract as risk-free and receives W if the user evaluate the contract as risky. Otherwise, if the contract proves to truthfully be *risky*, then the user receives R if the user evaluated the contract as risky and W if the user evaluated the contract as risk-free.

4.6.3   Understanding the Provider Perspective

Let us try to examine the cloud provider perspective. The cloud provider will publish a contract, which is either *risk-free* or *risky*. However, eventually Nature will decide whether or not something happens to render the contract untrue. The cloud provider is aware of this probability and is willing to risk preparing a more risky or less risky contract for the cloud user. Consider the sequence of decisions for the game: there are two decisions that the cloud provider can make, to offer a risk-free (which in reality is simply less risky) or a risky contract to the cloud user. In turn, *Nature*, given its probability to affect the game, will support conditions that will make the risk true or untrue. In terms of payoffs to the cloud provider, there is a small payoff S for offering a risk-free contract, since this will result in a satisfied user and better reputation, and there will be a larger payoff L for convincing the user that the contract is risk-free, even if it is not, such that $P_{prov.} = L > S$.

4.6.4   The Extensive Form of the Game

The interaction between a cloud user and a cloud provider as described above can be modelled as a game with an extensive form. The game will be characterised by incomplete information because not all the *players* know the same information. *Nature* will play first to decide whether the risk is actually true or untrue. This play will not be revealed to the players until time has allowed for the revelation of this play or at least the perception of *Nature's* play from the user perspective. Figure 4.2 begins with Nature's play followed by the cloud provider's play and then the user's play as points in time. However, the user does not have the information on the previous moves made by the provider and by *Nature* at decision time. Similarly, the provider is not aware of *Nature's move* at decision time.

Even if the cloud provider and the cloud user will not receive information about Nature's play until later in the game, they both know the probability for each of *Nature's* moves. This probability is known ahead of time and may consider it in making their decision. Then the cloud provider needs to decide on the type of contract to present to the cloud user. Therefore, the next move after *Nature's* move is the move of the cloud provider. The final move is by the cloud user, who makes a decision on whether or not to trust the contract or not. The illustration of the decisions by the players indicate the sequence of moves. It is important to recognise the

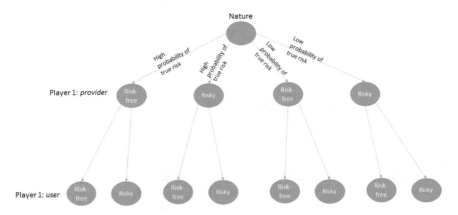

Fig. 4.2 Interaction between provider and user as a sequential moves model

incomplete information by the cloud user, who does not know whether to trust the cloud provider or not, and needs to make a decision that will eventually affect the QoE.

4.6.5   *The Normal Form of the Game*

To better analyse this model, we will convert it to a normal form game (i.e. we will not consider the element of time and how the *players'* moves are, in fact, sequential).

The possible outcomes of the play are based on the player strategies and actions, which are based on players' behaviours. We can first identify the types of behaviours for the normal form game, and we list them below. We refer to decision-making entities as the players, and we have two players in this game: the cloud provider and the cloud user.

The cloud provider has four types behaviour, T_i, $i \epsilon \{1, 2, 3, 4\}$:

1. If Nature plays *True Risk*, offer a risky contract, if Nature plays *Untrue Risk*, offer a risky contract (i.e. always offer a risky contract).
2. If Nature plays *True Risk*, offer a risky contract, if Nature plays *Untrue Risk*, offer a risk-free contract (i.e. always offer Nature-aligned contract).
3. If Nature plays *True Risk*, offer a risk-free contract, if Nature plays *Untrue Risk*, offer a risky contract (i.e. always offer a non-Nature-aligned contract).
4. If Nature plays *True Risk*, offer a risk-free contract, if Nature plays *Untrue Risk*, offer a risk-free contract (i.e. always offer a risk-free contract).

The cloud user also has two distinct types of behaviour, T_j, $j \epsilon \{1, 2, 3, 4\}$:

1. Evaluate the contract as risk-free, when Nature plays *Untrue Risk*.
2. Evaluate the contract as risky, when Nature plays *True Risk*.
3. Evaluate the contract as risk-free, when Nature plays *True Risk*.
4. Evaluate the contract as risky, when Nature plays *Untrue Risk*.

The behaviour types affect the strategies of the two players characterised as S_i for the cloud provider and S_j for the cloud user. The strategies, in turn, affect the payoffs of the two players from this interaction. Let the user be player 1 and the provider be player 2, such that payoff $P_1(S_i) = L > S$ and payoff $P_2(S_j) = R > W$.

To be able to easily compare the payoffs for the two players, we can review the strategies according to the possible outcomes described above. The strategies are listed next:

Player 1, i.e. the cloud provider has four strategies, or types of actions, S_i, $i \epsilon \{1, 2, 3, 4\}$:

1. Nature plays *True Risk*, and Player 1 offers a risk-free contract.
2. Nature plays *True Risk*, and Player 1 offers a risky contract.
3. Nature plays *Untrue Risk*, and Player 1 offers a risk-free contract.
4. Nature plays *Untrue Risk*, and Player 1 offers a risky contract.

Player 2, i.e. the cloud user also has four strategies or types of actions, S_j, $j \epsilon \{1,2,3,4\}$:

1. Evaluate the contract as risk-free and verify that it is risk-free (Nature plays *Untrue Risk*).
2. Evaluate the contract as risky and verify that it is risky (Nature plays *True Risk*).
3. Evaluate the contract as risk-free and discover that it is risky (Nature plays *True Risk*).
4. Evaluate the contract as risky and discover that it is risk-free (Nature plays Untrue Risk).

For simplicity, let us give some values to the payoff variables S, L, R, W and try to analyse an example of this interaction. Let $S = 1$ and $L = 2$ for player 1, i.e. the cloud provider. Let $R = 1$ and $W = 0$ for player 2, i.e. the cloud user. Remember that in terms of payoffs to the Player 1, there is a small payoff S for offering a risk-free contract, since this will result in a satisfied Player 2 and better reputation for Player 1, and there will be a larger payoff L for convincing Player 2 that the contract is risk-free, even if it is not. Moreover, for Player 2, in case the contract proves to truthfully be *risk-free*, Player 2 receives R if Player 2 evaluated the contract as risk-free, and receives W if the evaluation of the contract was one of it being risky. Otherwise, if the contract proves to truthfully be *risky*, then Player 2 receives R if Player 2 evaluated the contract as risky and W if the evaluation of the contract was one of it being risk-free.

Next, the payoffs are presented in a tabular form (Table 4.1). Player 1 strategies are labelled across the rows of the table and Player 2 strategies are labelled across the columns of this table. Each cell presents the specific payoffs to player 1 and player 2, in a set of numbers depicted as Player 1 and Player 2 for a set of strategies of the corresponding row and column where the cell is located. Note that due to the sequential nature of the game, i.e. the fact that the time element will reveal the play of Nature to the players, the playoffs are revealed considering the play of Nature is either *True Risk* or *Untrue Risk*; that is why some of the table cells do not contain a set of payoffs. This is a result of incomplete information, i.e. there are a number of strategies but not all are available in terms of payoff after Nature's play.

For S_i, $i = 1$ and S_j, $j = 2$, Nature plays *True Risk,* but the contract designed was risk-free, and not affected by Nature, so Player 1 gets 1 point for designing a risk-free contract, but Player 2 does not get any payoff because Player 2 evaluated the contract as risky.

Table 4.1 Payoffs from the various sets of player strategies

	$S_j, j = 1$	$S_j, j = 2$	$S_j, j = 3$	$S_j, j = 4$
$S_i, i = 1$		(1,0)	(3,1)	
$S_i, i = 2$		(0,1)	(2,0)	
$S_i, i = 3$	(3,1)			(1,0)
$S_i, i = 4$	(2,1)			(0,0)

For S_i, $i = 1$ and S_j, $j = 3$, Nature plays *True Risk,* but the contract designed was risk-free, and not affected by Nature, so Player 1 gets 1 point for designing a risk-free contract and 2 points for convincing Player 2 that it is risk-free, i.e. a total of 3 points. Player 2 gets 1 point from this interaction, because Player 2 correctly evaluated the contract as risk-free.

For S_i, $i = 2$ and S_j, $j = 2$, Nature plays *True Risk,* and the contract designed was risky. Player 1 does not get a point from designing a risk-free contract but gets the 2 points from convincing Player 2 that it is a risk-free contract. Indeed, Player 2 evaluates the contract as risk-free, but since Nature shows evidence of it being a risky contract, Player 2 receives 0 points for being wrong in the initial evaluation of the proposed contract.

For S_i, $i = 2$ and S_j, $j = 3$, Nature plays *True Risk,* and the contract designed was risky. Player 1 does not get a point from designing a risk-free contract. Player 1 does not get the 2 points from convincing Player 2 that it is a risk-free contract either, so receives 0 points. Player 2 evaluates the contract as risky and since Nature also supported this, Player 2 receives evidence that the contract was indeed risky and receives 1 point for being right.

For S_i, $i = 3$ and S_j, $j = 1$, Nature plays *Untrue Risk,* and the contract designed was risk-free. Therefore, Player 1 receives 1 point for designing a risk-free contract. Player 1 does not convince Player 2 that it is risk-free, so does not receive any more points. Player 2 evaluates the contract as risky but since Nature does not provide any evidence that the contract is risky, Player 2 receives 0 points for being wrong.

For S_i, $i = 3$ and S_j, $j = 4$, Nature plays *Untrue Risk,* and the contract designed was risk-free. Therefore, Player 1 receives 1 point for designing a risk-free contract. Player 1 also manages to convince Player 2 that the contract is risk-free, so Player 1 receives two additional points, i.e. 3 total points. Player 2 evaluates the contract as risk-free, and since Nature does not provide any evidence that the contract is risky, Player 2 receives 1 point for being right.

For S_i, $i = 4$ and S_j, $j = 1$, Nature plays *Untrue Risk,* and the contract designed was risky, but not affected by Nature, so Player 1 does not receive 1 point for designing a risk-free contract. Player 1 receives 2 points for convincing Player 2 that this is a risk-free contract. Player 2 evaluates the contract as risk-free, and since Nature does not provide any evidence that the contract was risky, Player 2 receives 1 point for being right.

For S_i, $i = 4$ and S_j, $j = 4$, Nature plays *Untrue Risk,* and the contract designed was risky, but not affected by Nature, so Player 1 does not receive 1 point for designing a risk-free contract. Player 1 does not receive 2 points for convincing Player 2 that this is a risk-free contract, since Player 2 evaluates the contract as risky. For Player 2, since Nature does not provide any evidence that the contract was indeed risky, the evaluation is not verified and Player 2 receives 0 points for being wrong.

As mentioned earlier in the chapter, it is clear that all payoffs for the cloud user consider a subjective view of the user, since the model evaluates the user payoff, which relates directly to user QoE and needs to be evaluated from the user's point of view.

4.6.6   A Closer Look at the Game Payoffs

The examples above, both the sequential game model as well as the normal form game model, present *Nature's* actions as final. The payoffs are given without the concern of time. However, in reality, there needs to be a time lapse before the players know whether or not they have succeeded, since there needs to be some malicious or hazardous activity that will activate the security clause of the user's contract. What is actually known or can be estimated from the beginning is the probability of *True* or *Untrue Risk*, given a variety of factors briefly mentioned at the beginning of this chapter. Thus, the payoffs are only true according to these probabilities.

Let the probability of *True Risk* be represented as p, and the probability of *Untrue Risk* be represented as $1-p$. Then the payoffs can be modified to include these probabilities as shown in Table 4.2. The variables p and $1-p$ can be replaced by appropriate probabilities such that:

$$p + (1-p) = 1, \quad \text{e.g. } p = 0.2, \quad \text{and} \quad 1 - p = 0.8.$$

By multiplying the probabilities, we end up with the Table 4.3.

Let us discuss what the above means. We will first investigate the case when the probability p is high, i.e. closer to 1, in which case the options in the first two rows will be activated. We will then investigate the case when the probability p is low, i.e. closer to 0, in which case the options in the last two rows will be activated.

In the first two rows, the probability of *True Risk* should be high, because these are the payoffs affected by evidence of malicious or hazardous activity as discussed by the explanation of the strategies themselves. Therefore, p is close to 1 and Player 1, i.e. the cloud provider has more to gain by providing a risk-free contract, no matter what Player 2 chooses, because by selecting a risky contract, there is the probability to be left with zero payoff. Therefore, Player 1 will select to play on the first row of Table 4.3 (given that p is high, which implies that we are only considering the first 2 rows of the table).

If Player 2 knows the probability of p is high, then when presented with the contract, Player 2 can assume that Player 1 will select row 1 to avoid a zero payoff and thus it is the best play for Player 2, to select column 3, to avoid zero payoff as well. Eventually, the best set of strategies for the two players is that Player 1 presents a risk-free contract and Player 2 evaluates it as a risk-free contract. In summary, if there is a high probability that opportunity will arise for risks to materialise, it is

Table 4.2 Payoffs from the various sets of player strategies, considering probabilities for Nature play

	$S_j, j = 1$	$S_j, j = 2$	$S_j, j = 3$	$S_j, j = 4$
$S_i, i = 1$		$(1,0) \cdot (p)$	$(3,1) \cdot (p)$	
$S_i, i = 2$		$(1,0) \cdot (p)$	$(2,0) \cdot (p)$	
$S_i, i = 3$	$(3,1) \cdot (1-p)$			$(1,0) \cdot (1-p)$
$S_i, i = 4$	$(2,1) \cdot (1-p)$			$(0,0) \cdot (1-p)$

Table 4.3 Payoffs with probabilities simplified

	$S_j, j = 1$	$S_j, j = 2$	$S_j, j = 3$	$S_j, j = 4$
$S_i, i = 1$		$(p, 0)$	$(3p, p)$	
$S_i, i = 2$		$(0, p)$	$(2p, 0)$	
$S_i, i = 3$	$(3 - 3p, 1 - p)$			$(1 - p, 0)$
$S_i, i = 4$	$(2 - 2p, 1 - p)$			$(0, 0)$

Table 4.4 When p is closer to one (1)

	$S_j, j = 2$	$S_j, j = 3$
$S_i, i = 1$	$(p, 0)$	$(3p, p)$
$S_i, i = 2$	$(0, p)$	$(2p, 0)$

better to have a risk-free contract by the cloud provider, and it is better for the cloud user to trust it.

In the last two rows, the probability of *True Risk* should be low, because these are the payoffs affected by lack of evidence of malicious or hazardous activity as discussed by the explanation of the strategies themselves. Therefore, p is close to 0 and Player 1, i.e. the cloud provider has more to gain by providing a risk-free contract, no matter what Player 2 chooses, because by selecting a risky contract, there is the probability to be left with zero payoff. Therefore, Player 1 will select to play on the third row of Table 4.3 (given that p is low, which implies that we are only considering the last two rows of the table).

If Player 2, knows the probability of p is low, then when presented with the contract, Player 2 can assume that Player 1 will select row 3 to avoid a zero payoff and thus it is the best play for Player 2, to select column 1, to avoid zero payoff also. Eventually, the best set of strategies for the two players is that Player 1 presents a risk-free contract and Player 2 evaluates it as a risk-free contract. In summary, if there is a low probability that opportunity will arise for risks to materialise, it is better to have a risk-free contract by the cloud provider, and it is better for the cloud user to trust it.

Table 4.4 depicts the best set of actions when p is closer to 1, and Table 4.5 depicts the best set of actions when p is closer to 0.

As a final remark, it is important to highlight the conclusion from the above payoff tables (Tables 4.4 and 4.5). What seems to prevail from this model is that no matter what the Nature play is, or, no matter how high (or low) the probability of *True Risk* is, both players would benefit more if they were to engage in a trustful interaction. In a trustful interaction, the cloud provider offers a risk-free contract, and the cloud user trusts the contract and evaluates it as risk-free.

Table 4.5 When p is closer to zero (0)

	$S_j, j = 1$	$S_j, j = 4$
$S_i, i = 3$	$(3 - 3p, 1 - p)$	$(1 - p, 0)$
$S_i, i = 4$	$(2 - 2p, 1 - p)$	$(0, 0)$

4.7   Discussion

Cloud computing is undoubtedly becoming increasingly popular, especially with business owners as an alternative to in-house technological infrastructures. Services offered by the cloud vary in terms of the level of control that the cloud customer has on system configuration, as well as in terms of the level of management that the cloud provider offers to the customer. In any case, service agreements are put in place and the cloud customer, or cloud user (as is the term we have used throughout the chapter), is asked to trust the cloud provider (and hence the proposed agreement) and enter into this interaction.

The user needs to make decisions based on incomplete information, such as the lack of knowledge on matters of network traffic, network and hardware technology, information about co-existing virtual machines on the same hypervisor, etc. On the other hand, the low cost, the elasticity of cloud resources and the array of available services are positively motivating factors for entering into the agreement.

In this sense, user experience can be controversial. On the one hand, it may turn out to be an easier and more comfortable experience than the in-house alternative. On the other hand, it can create a sense of lack of control in terms of privacy or personal data protection, with the worst-case scenario being that concerns about lack of control being verified. The risks do not only refer to malicious behaviour, which is an important risk, but even to natural disasters or system failures. Even if the customer trusts that the cloud provider can secure the system against any malicious attack, a natural disaster could cause harm to the provider's infrastructure without the user being able to do anything to protect its own virtual machines or data. But even without looking into the probability of natural disasters, the lack of control can be translated into probable data loss or data corruption due to system reasons.

From the cloud provider's perspective, it is important to gain a good reputation and keep it so that cloud users can trust the cloud services, even in the presence of generic risks. By ensuring that the users feel secure, in time their trust is gained and long-term positive relationships between cloud providers and cloud users are profitable for the providers. Remember that the pricing model of using cloud services is a *pay-as-you-go* model.

The normal form game presented above shows the decisions by both the provider and the user as single, simultaneous decisions, which result in payoffs based on the decisions of the players, but also on the play of *Nature*, i.e. whether the risk of disaster or attack is realised. The model holds that a robust set of security controls by the cloud provider can help systems recover, even in the case of a negative *Nature* play. In fact, representatives of main cloud service providers admit that cloud services have in place a high number of security controls to ensure that customers will remain trustful of the interaction with the cloud platform [10].

On the other hand, the cloud user has some decisions to make as well. Our model does not assume that the cloud user has any knowledge of combating risk when using the cloud, but in reality, especially for a business owner, it is important to have a knowledgeable local IT manager or IT staff. Once the local support is in place, then the user can select a cloud service provider to interact with, preferably a reputable cloud provider, offering services of good quality and strong security support.

References

1. Antoniou J (2019) Using game theory to address new security risks in the IoT. In: 'Game Theory, the Internet of Things and 5G Networks', part of the: 'EAI/Springer Innovations in Communication and Computing' book series, Springer, https://doi.org/10.1007/978-3-030-16844-5_2

2. Bello SA, Wakil GA (2014) Flexible pricing models for cloud services. Tans Netw Commun 2(5). https://doi.org/10.14738/tnc.25.281

3. Chang, V., Bacigalupo, D., Wills, G., De Roure, D. (2010). A categorisation of cloud computing business models. In Proceedings of the 10th IEEE/ACM International Conference on Cluster, Cloud and Grid Computing, Melbourne, VIC, 2010, pp 509–512

4. Chang V, Walters RJ, Wills G (2013) The development that leads to the cloud computing business framework. Int J Inf Manag 33(3):524–538. ISSN 0268-4012, https://doi.org/10.1016/j.ijinfomgt.2013.01.005

5. Hashizume K, Rosado DG, Fernandez-Medina E, Fernandez EB (2013) An analysis of security issues for cloud computing. J Internet Serv Appl 4(5) Springer, https://doi.org/10.1186/1869-0238-4-5

6. McLelland R, Hackett Y, Hurley G, Collins D (2014) Agreements between cloud service providers and their clients: a review of contract terms. In Proceedings of Arxius i Industries Culturals conference, Girona, October 2014

7. Hsu P-F, Ray S, Li-Hsieh Y-Y (2014) Examining cloud computing adoption intention, pricing mechanism, and deployment model. Int J Inf Manag 34(4):474–488. ISSN 0268-4012

8. Perez-Botero D, Szefer J, Lee RB (2013) Characterizing hypervisor vulnerabilities in cloud computing servers. Proceedings of the 2013 International Workshop on security in Cloud Computing, pp 3–10, https://doi.org/10.1145/2484402.2484406

9. Szefer J, Keller E, Lee RB (2011) Eliminating the hypervisor attack surface for a more secure Cloud. Proceedings of the 18th ACM Conference on Computer and Communications Security, pp 401–412., https://doi.org/10.1145/2046707.2046754

10. Wall M (2016) Can we Trust Cloud Providers to keep our data safe?, Technology of Business, BBC News, published on 29 April 2016. [Online] https://www.bbc.com/news/business-36151754

Chapter 5
Modern Technology Interfaces: Usability, Sociability Challenges and Decision-Making

5.1 Introduction

The chapter investigates how decision-making, relevant to modern interface design, can improve user experience by approaching user interface design from a user experience perspective.

We will explore digital interfaces that allow for interactions between different users, and the design should consider the element of user-to-user communication as part of the experience. The type of digital interfaces we will look at are those that allow for knowledge exchange between users, e.g. mentoring platforms. The design of the interface can play an important role in improving the user experience. When referring to experience in relation to a specific interface, often with the use of the acronym UX, we generally refer to elements such as interface usability and design features. To achieve improved UX, designers need to consider elements of user experience as part of the design process. This is already widely done in interaction design and the specific design flavour is referred to as *user experience design* or *UX design*.

As mentioned above, the chapter explores the interaction of multiple users through digital interfaces that focus on knowledge exchange. In particular, the chapter investigates a specific example that considers two types of users that we will refer to as *mentors* and *mentees*, on a knowledge exchange type of an interface. Specifically, we are concerned with a mentoring platform, where there are two types of users, the *mentors* that delivery the knowledge and the mentees that receive the knowledge. Prior to looking at specific considerations for designing such an interface, we will discuss the area of UX design in general.

© Springer Nature Switzerland AG 2021   55
J. Antoniou, *Quality of Experience and Learning in Information Systems*,
EAI/Springer Innovations in Communication and Computing,
https://doi.org/10.1007/978-3-030-52559-0_5

5.2   User Experience Design

UX design considers digital interfaces and how such interfaces are designed around the user experience itself. Even though, in this chapter, we will try to explore some group dynamics of interacting through digital interfaces, UX design addresses all kinds of digital interfaces and various types of user audiences through UX design techniques that can help to improve the user experience. UX design is a collection of techniques that help the interface designer approach the design process from the perspective of user experience rather than from the perspective of the interface functionality and the functionality of its particular elements. This is a challenging process, because this user-centred approach needs to move away from considering the technical aspects of the various interface components or how well these components fit their purpose as interface tools. Instead, UX design needs to move towards a more *holistic* understanding of user experience. Once this is achieved, the next step is to define how the design process itself can be enriched in order to understand and improve the user experience from the design point of view.

When approaching design in a more holistic manner, we need to identify the specific elements or aspects of the design that may directly affect user experience. These must be examined in order to identify how they can actually contribute to the interface design. We will highlight five elements identified as the main aspects of experience that can be mapped to design elements. We do not claim this to be an exhaustive list of elements; the interested reader can revisit this list and update it in order to better fit specific interfaces.

The first element that can affect user experience is *usefulness* and how an interface component or the interface itself is useful to the user audience. A useful interface should also be *clear* in terms of its purpose. The second element is *usability*, and how easy an interface component or the whole interface itself is to use, in terms of *navigating* and *interacting* with it. There should also be a consideration of whether the targeted user audience would need additional guidance needed to be able to use the interface. A third element is therefore, *attainability*, which is the answer to the question of whether the interface components and the interface as a whole are *easy to master* with minimal instructions. A fourth element to consider for user experience design is the actual *aesthetics*, as with traditional interface design, i.e. whether the interface's visual appearance and its overall design are appealing to the specific user audience. Finally, the last element to consider is the *emotional impact* of the interface. The interface is expected to evoke emotions from the user audience in response to the interface itself and the product brand, so it is important that user experience design ensures that these emotions are *positive*, and if possible that they have a lasting positive impact on the targeted user audience.

5.3   Emotional Design

The idea of emotional design, and ways in which positive emotions can be inspired or motivated by interacting with a digital interface, is very interesting. According to recent studies, evoking emotions from the users from an interaction with a visual interface can be approached by considering the different ways in which people react to it [3]. In fact, people react in three different ways when it comes to emotions related to such an interactive, visual experience: visceral, behavioural and reflective. If all aspects of their reaction, i.e. the visceral, the behavioural and the reflective, are positive, then the user is left with a lasting positive user experience.

There are several aspects that need to be considered by the designers, as we mentioned earlier in the chapter, in addition to emotional design, for example how desirable is the product itself, or whether there can be a trusting relationship with the application provider. Such elements may, positively or negatively, affect the experience and the level that this experience can evoke positive emotions.

Nevertheless, assuming that the product is desirable and trustworthy, we proceed to discuss the emotional design aspects according to the visceral, behavioural and reflective reactions of the user audience. A *visceral* reaction usually represents the initial sensory experience when interacting with a specific interface. A positive reaction may be a result of well-thought interface aesthetics or an effective and inclusive design. This is in fact the first impression of a user and represents the product's initial attractiveness to a specific user audience.

A *behavioural* reaction represents the user's feelings once some interaction with the product has occurred. Design aspects that can improve the behavioural reaction include usability and performance. For achieving this, the designer must approach the design by understanding the user needs and expectations in order to ensure that the product provides an easy, useful, effective and efficient solution to the user-centred design objectives. While the positive visceral reaction will attract the user to try out the product, a positive behavioural reaction will encourage the user to use the product repeatedly.

A *reflective* reaction deals with the feeling of the user after the experience. How does the user remember the experience, if time is allowed to pass? Whether the user will think about the product or talk about it with others are testaments to a positive reflective experience. This aspect of the design is not directly linked to the interface itself, or its functionality, but with the overall concept, its significance and its wider impact (cultural, professional, societal, etc.).

New types of products and additional needs on interfaces make the design even more challenging. For example, if we consider a knowledge exchange platform between different types of users, who not only interact with the interface but also with each other, then the simple considerations on functionality and design that a non-collaborative interface usually tackles become less significant. Instead, focus is placed on considerations of how communication and interaction between users is achieved, and what the behavioural aspect of emotional design means, when alternate preferences may need to be satisfied.

5.4 Sociability in Design

When considering behavioural reactions from the use of an interface by a group (or groups) of users, the primary objective is to achieve usability for the group(s). This is often referred to as *sociability*, since it implies intra- and inter-group interaction and communication. Therefore, the designer needs to consider the types of relationships that are expected or allowed between the users. Hence, the actions are enabled accordingly and digital journeys for the users are outlined and tested. Next, the designer must consider the ways these actions may affect experience, e.g. is navigation and selection easily mastered, or, is the expected interaction familiar?

The success of such an interface is directly linked with the users' collective experience, more so than any other aspect that affects the emotional aspect of the design. The aesthetic aspect or initial attractiveness of the interface carries less weight than the communication component. The effectiveness and efficiency of this communication is what will attract the user to come back, especially if the rest of the group returns to the interface. Thus, the reflective reaction also carries less weight because it is less about individual preference or word of mouth, but more of a collective decision between the group(s) and their successful communication. So, the idea of group usability or sociability [4] should be evaluated as part of determining a successful UX design, based on the behavioural aspect of the emotional element of the interface design.

A mentoring platform is the considered example for this chapter. A mentoring platform is a knowledge exchange platform between two types of users, usually referred to as the *mentors* and the *mentees*. Each *mentee* can seek mentoring from one or more *mentors* to achieve a specific goal or to learn a specific piece of knowledge. Knowledge exchange represents a form of informal education, through which knowledge is shared. When we refer to a specific knowledge exchange platform, such as a mentoring platform, we are in essence, referring to a systematic approach to sharing knowledge that people acquire through their experience. For mentoring platforms in particular, experience refers to professional and work experience, including successes and challenges in order to support less experienced professionals gain knowledge that is essential for professional success.

Usually, the purpose of knowledge exchange is to connect practitioners that work in the same field or professionals of a particular discipline with each other. Knowledge exchange can occur through discussion, presentations, question and answer sessions, etc., so that they may learn from one another and improve their knowledge and professional skills. Sharing knowledge, especially experiential knowledge, is a key ingredient in innovation. Moreover, according to UNICEF [5], knowledge exchange is a key element in achieving learning from experience and applies that learning to improve actual work results.

To achieve these objectives, knowledge exchange makes use of specific tools and approaches, including face-to-face exchange, either through physical interaction or through virtual chatrooms and real-time collaborative software tools. The approach that we will consider in our example scenario in this chapter is that of a software platform that facilitates online networking and knowledge exchange. This is helpful as a tool because it considers a group of users that undertake different mentoring roles, who can work together across geographic and organisational barriers. Such mentoring platforms are helpful as they support professional communities to grow, by sharing resources and eventually co-creating new knowledge. Overall, such platforms function as a means to forming new partnerships between professionals and at the same time provide a space for these new effective professional networks to interact, seek or offer professional advice, learn and train. Through the knowledge exchange or mentoring platform, both time and money can potentially be saved in dealing with professional challenges.

5.5 Designing for Mentors and Mentees

Learning is often a big component and motivational factor for such mentoring platforms. The idea of the interface functionality is to simplify learning by offering a non-formal type of education to non-traditional students, e.g. professionals that aim to improve their knowledge, practitioners that want to continue lifelong learning, novices that want to acquire expertise in a specific area and many other examples. A knowledge exchange digital platform provides a collaborative digital interface that hosts a mentoring scheme such that *mentors*, i.e. the users that will provide the knowledge, can support *mentees*, i.e. users that want to receive the knowledge. We will not be concerned with the design aesthetics, but the scenarios that we will explore will deal with the expected behaviour of the users, as a component of emotional design, as part of UX design.

Since each *mentee* can seek mentoring from one or more *mentors* to achieve a specific learning objective, there needs to be a decision on how the best *mentor* is matched to a specific *mentee*. We propose that the *mentee* selects a *mentor*, out of the *mentors* available to deliver a specific learning objective. Since this is a peer-to-peer example platform, there is no central authority that can guide this selection decision, e.g. by rating the quality of the *mentors* across the offered learning objectives.

Therefore, a design option would be to allow for peer-to-peer *positive rating* of the *mentors*, avoiding negative rating or comments, but enabling a historical insight into the past teaching attempts of specific *mentors*. By adding the option to positively rate a *mentor's* delivery, a globally available *rating* can inform a *mentee's* selection of *mentor*, because the rating that reflects past *mentor* behaviour, e.g. quality delivery, achieved learning objectives, etc. At this point, it is important to mention that one of the initial assumptions is that the *mentors* themselves will have strong incentives for participating on the platform, given that the incentive for the *mentees* is the actual learning. Nevertheless, avoiding the use of negative ratings or comments can serve as an additional incentive to the *mentors* as well.

We will not look into a specific type of mentoring platform, i.e. we will not investigate the specific type of knowledge that will be exchanged. Therefore, we will not speculate as to what the content-specific incentives can be, for either the *mentors* or the *mentees,* as they may change for different types of content. However, we will take for granted that such incentives exist. The interested reader can find specific examples for different types of mentoring platforms in peer-reviewed literature [6, 7].

Based on the assumptions and objectives of both groups of users, we will next outline a model of interaction between one *mentee* selecting between a number of *mentors* based on a simple rating scheme that easily adapts to the *mentors'* past mentoring success. The purpose of the model is to discuss how a satisfying user experience can be achieved through this simple mechanism, i.e. positive rating mechanism, by breaking down the decisions taken during a simple selection of the preferred *mentor* or *mentors* by a *mentee* for a specific learning objective, and the subsequent interaction during the delivery of the learning objective.

5.6   Mentor Selection Model

The model examines the relationship between a *mentee* and a number of *mentors*, when all entities actively participate in an interactive mentoring platform and the *mentors* can satisfy a specific learning objective required by the *mentee*. The *mentee* is thus required to select a *mentor,* or a group of *mentors*, based on the learning requirements. The proposed model describes the mechanism of *mentor selection* in a *mentee-mentors* interaction on a knowledge exchange interface. The model is based on the evaluation of a utility function by the *mentee,* which reflects the subjective evaluation of the *mentors* that want to participate in the interaction. The idea of using a utility function to quantitatively evaluate the expected experience from an interaction has been used in technology-related selection mechanisms in literature [1, 2]. Following the selection model, the chapter will take a closer look at the interaction between the *mentee* and the selected *mentor,* and provide more details as to the use and effect of the positive rating feature, introduced by the design.

The mechanism explored next deals with the specific decisions and information (e.g. rating) through which the *mentee* selects the best *mentor* with which to undertake the required learning. The *mentor selection* decision is modelled as a *game* between one *mentee* participating on the knowledge exchange platform and the participating *mentors* that are available to the specific *mentee* (and are relevant to the interaction, i.e. they match the specific learning objective requirements).

In the model of the interaction, we approach the modelling of the *mentors* as one single *player,* in order to allow the *mentor* participants to strategise simultaneously. *Consequently,* the payoff for the *player* referred to as *mentors* is given as an array of payoffs corresponding to each of the individual *mentors* available.

The situation we model is the following:

The *mentee* has the first move and offers an incentive, e.g. a compensation, reputation incentive, etc. to *mentors* for participating in a particular learning activity. This may be trivial in some mentoring schemes, i.e. the *mentors* will always be expected to accept the incentive. For the purposes of the model, we proceed to allow the *mentors* to examine the incentive and make a decision concerning whether to accept or reject the offer. Then the *mentee* is informed about how many of them have accepted and how many have rejected the incentive.

In reality, since this is a peer-to-peer model, then there is no central authority to enforce participation of any entity, but it is expected that there are incentives in place. Even though, the assumption is that all *mentors* have incentives to participate in the knowledge exchange platform, this does not necessarily mean that all *mentors* share the same incentives. Therefore, any subset of *mentors* could accept or reject the proposed offer for a specific type of delivery, including the case that all *mentors* accept it, and the case that all *mentors* reject it. In the latter case, the interaction terminates, and hence the *game* terminates. If the game terminates without a selection, then both the *mentee* and the *mentors* receive zero payoffs.

If one or more *mentors* accept to participate in the learning interaction, then for each *mentor*, the *mentee* evaluates an internal utility, in terms of *expected UX*, $UX_i(r)$, for *mentor i* with rating r. The *mentee* evaluates the resulting output of the utility function for each *mentor i* to identify the *mentor* with the higher utility. The user, in turn, selects the *mentor* that is predicted to offer the highest utility output. If only one *mentor* accepts, the selection is trivial.

The evaluation of $UX_i(r)$ is mainly based on r, the overall rating that each mentor accumulates from participating on the mentoring platform. The peer-to-peer positive rating is evaluated through the interface functionality. This will be explored later in the chapter as part of the interaction strategies that follow the *mentor selection* process. In addition to the positive rating, the utility function also considers possible negative costs that may result from the interaction with a specific *mentor*, e.g. time duration of the knowledge exchange activity, additional resources needed, number of prerequisites, etc. Therefore, the overall utility ends up being the difference between positive rating and potential cost. Note that both the gain and the cost are represented as a fraction between zero (0) and one (1). Because of the subjective evaluation of utility, it is expected that the resulting utility output will be different for each *mentor* as evaluated by a particular *mentee*.

The *mentee's* decision to select one of the *mentors* is followed by the decision of the specific *mentor* to start interacting with the *mentee*, using the knowledge exchange interface.

It is crucial to follow this interaction through, because at the end of this interaction, the *mentor* will end up with an updated overall positive rating that will reflect this interaction. Specifically, an improved positive rating will show that the learning interaction went well, whereas a decreased positive rating will demonstrate the opposite. Therefore, we need to understand potential strategies that may result in this rating. From a design perspective, different strategies will help us explore how the overall experience of the interacting entities changes as a result of the rating feature. The aim is to guarantee, to a reasonable extent, that an improved user experience can result from this for both the *mentor* and the *mentee*. The following sections will follow this interaction through.

Before moving on to the *mentor-mentee interaction* model, a final remark on the *mentor selection* game model must be made. The *mentors* that are not selected by a *mentee* for a specific delivery will end up with zero (0) payoffs (as a result of the specific *mentee* request). The payoffs for the *mentee* and the selected *mentor(s)* can be calculated based on the interaction strategies during the interaction. A detailed model of the *mentor-mentee interaction* is elaborated next. Finally, the sequence of decisions and potential moves of the *players* in the *mentor selection* game is illustrated in Fig. 5.1.

5.7   Mentor-Mentee Interaction Model

Once the *mentee* has selected the *mentor* to interact with, the *mentor* is responsible to deliver the learning. The *mentor* can deliver the promised learning at the required quality or try to cheat out of this responsibility by lowering the required quality of the learning delivered. The motivation for such *cheating* behaviour is often the reduction of cost for the *mentor.* The cost may be evaluated by the *mentor,* as a combination of factors, such as planning effort, time spent, resources used, etc. Therefore, a cheating behaviour could result in the *mentor* spending less time or using less resources for instance.

This is tricky for participants in a peer-to-peer knowledge exchange platform, because there is no central authority to monitor the learning interaction. The lack of monitoring by a central authority can end up being risky for the quality of the

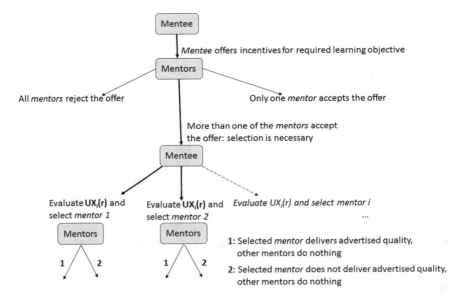

Fig. 5.1 Mentor selection model

learning since *mentors* may want to put less effort than promised into the delivery or share less resources than advertised with the *mentees*. If such cheating behaviour becomes frequent, the *mentees* will lose trust in the quality of learning received through the knowledge exchange platform, and consequently lose motivation to participate. This is the reason that the positive rating feature is very important to ensure that the interaction quality and overall learning is somehow monitored or recorded.

Let us characterise the action of delivering quality learning by a *mentor* as a cooperative action, and, let us characterise the attempt of a *mentor* to cheat out of a quality delivery of the learning as a non-cooperative action. Thus, the decisions that a *mentor* can make include: (i) the decision to cooperate with the *mentee* and (ii) the decision to defect from cooperation, while the rest of the *mentors* do not interact any further with the *mentee* during the specific learning delivery. From then on, the interaction between the *mentee* and the *mentor* will change based on the *mentor's* decision on whether to follow a cooperative or a non-cooperative strategic behaviour. The *mentee* can respond to a non-cooperative decision by leaving the interaction with the specific *mentor* or by not rating the mentor positively.

We will primarily examine the interaction by considering that the *mentor's* decision is a once-off decision, i.e. that it is not affected by outcomes of previous interactions and that it does not affect any future interactions between the *mentor* and the *mentee*.

Consider that the *mentee* offers the incentive λ, a positive reward, which is accepted by the *mentor,* who needs to make a decision of whether to offer the quality delivery or not to offer it and save costs. Let the cost of offering quality delivery be κ and the cost of reducing quality be κ', where $\kappa > \kappa'$. Thus, a *mentor's* utility UX_j, where j represents the particular *mentee*, can be one of:

$$UX_j = \lambda - \kappa, \quad \text{or}$$
$$UX_j = \lambda - \kappa', \quad \text{where } \kappa > \kappa'.$$

Based on the *mentor's* decision, the *mentee* can either receive full value v from the delivery, or decreased value from the delivery, v', where $v > v'$. Thus, in terms of payoff, the *mentee* receives one of two utilities:

$$UX_i = rv - c, \text{or,}$$
$$UX_i = rv' - c, \text{where } v > v'.$$

The overall utility is affected by the overall positive rating r, which should be a high value, since the specific *mentor* was selected by the *mentee* the selection was mainly based on the value of positive rating given by the digital interface for the participating *mentors* that were relevant to the specific learning objective.

Given the *mentor's* choice of actions, both players aim to maximise their payoffs. At this stage, the *mentee* only has one move, to participate, because the lack of positive rating, which will take place after the interaction has ended, will not affect the game. The reason is that we have made the initial assumption that any decisions made during the one-shot game, do not affect any future interactions between the two players. Of course, the rating will end up affecting the *mentor's* behaviour, but we will explore this further in the repeated form of this model. For now, we will analyse the moves in the normal form, i.e. the one-shot game. A final remark on the *mentee's* choice, that we should make, is that there is always the option of not participating in the interaction. If this decision is made by the *mentee* ahead of time, then the payoffs for both players will be zero (0).

Table 5.1 illustrates the possible actions and corresponding payoffs corresponding to each set of actions for the *mentee* and the *mentor* in the *mentor-mentee* interaction game. The table rows reflect the *mentor* actions and payoffs, and the columns reflect the *mentee* actions and payoffs. The payoffs are represented in the format *row payoff, column payoff*.

Consider that $\kappa < \lambda$ and $c < (rv')$, such that the payoff from interacting is always greater than zero (0). This is an assumption that is rational, because neither of the two entities would not accept an interaction without the possibility of a positive payoff. Therefore, we can see from Table 5.1 that both users gain more by interacting than by not interacting. The *mentor* knows that the *mentee* will prefer to interact and thus selects to cheat, because the payoff is higher in that case. Given the dominant strategies, the equilibrium of the game is for the *mentee* to interact and for the *mentor* to cheat. Given this preliminary analysis, the one-shot model will always motivate a cheating behaviour by the *mentor*.

The reality, however, is that such interactions between *mentors* and *mentees* are commonly not one-shot but reoccurring. A finite number of *mentors* and *mentees* participate on the mentoring, knowledge exchange platform and will continuously aim to interact, as long as they are participants, because that is the reason they are participating on the platform in the first place. In such interactions, i.e. recurring, the participating entities do not only seek the immediate maximisation of payoffs but instead the long-term optimal solution, such that the experience remains positive even after a number of repeated interactions. Often, mathematical models of repeated games are used in game theory to model such situations.

5.8 Repeated Interaction Model

There are two kinds of repeated game models: the finite horizon repeated game model and the infinite horizon repeated game model. The infinite horizon repeated game model is actually modelling games of unknown length. This kind of game model, i.e. the infinite-horizon game, does not imply that two particular entities will

Table 5.1 Mentor-mentee interaction – one-shot game model

	Mentee interacts	*Mentee* does not interact
Mentor cooperates	$(\lambda - \kappa, rv - c)$	$(0, 0)$
Mentor cheats	$(\lambda - \kappa', rv' - c)$	$(0, 0)$

keep interacting forever. It only means that we do not know how many times an interaction between two entities will take place. Since a *mentee* can request an interaction with a particular *mentor* for various learning objectives, then the number of total interactions is unknown at the time of the first interaction. Thus, we can categorise the *mentor-mentee* interaction model as an infinite horizon repeated game (since the *mentees* may keep requesting new knowledge exchange sessions from the *mentors*, but the number of such requests is not known).

A repeated game makes it possible for the players to condition their moves on the complete, previous history of the various stages of the interaction, by employing strategies that define appropriate actions for each period. Such strategies are called trigger strategies [8]. A trigger strategy is a strategy that changes in the presence of a predefined trigger; it dictates that a player must follow one strategy until a certain condition is met and then follow a different strategy, for the rest of the game.

One of the most popular trigger strategies is the grim trigger strategy [8], which dictates that the player participates in the relationship in a cooperative manner, but if dissatisfied for some known reason, leaves the relationship forever. The grim trigger strategy may be used by the *mentee* in the *mentor-mentee* interaction game, such that if the *mentee* is not satisfied in one interaction, i.e. the *mentor* does not provide the quality promised, then in the next interaction the *mentee* may punish the *mentor* by leaving the relationship forever. Leaving the relationship forever implies that the *mentee* will stop interacting with the specific *mentor* for subsequent requests of specific learning objectives.

Given such a strategy, the *mentor* has a stronger incentive to cooperate and provide the quality promised, since it faces the threat of being banned from interacting with the specific *mentee* permanently. Again, we need to remind the reader that both the *mentor* and the *mentee* want to stay active and participate in as many interactions as possible on the knowledge exchange platform.

The threat of non-renewal of the *mentee's* relationship with the *mentor* secures compliance of the *mentor* when we assume that the *mentor* is always making *rational* decisions (decisions that result in the highest payoff). Therefore, the *mentor* is motivated to satisfy the *mentee's* expectations. Exchanges based on such threats of non-renewing a relationship, which is based on a particular agreement between the two parties, i.e. deliver an advertised quality of learning, are often referred to as contingent renewal exchanges.

Therefore, the *mentee* can decide to employ a grim trigger strategy to elicit performance from the *mentor*. The loss of the relationship is costly to the *mentor* because it has a negative impact on the *mentor-mentee* relationship. According to the *mentee's* strategy, if the *mentee* is not satisfied with the provided quality, the *mentee* leaves the relationship. Hence, the *mentor* loses an entity to interact with, out of the finite number of participating entities on the knowledge exchange platform.

Another popular strategy used to elicit cooperative performance from an interacting counterpart, is for an interacting entity to follow a strategy that mimics the actions of the interacting entity. This strategy will give the interactive entity the incentive to play cooperatively, since in this way The reward will be a similar mirroring behaviour from the interacting counterpart. This strategy is referred to as tit-for-tat strategy [8]. Note that in a repeated-game model, it is possible to know the move of the opponent after each interaction, since the decisions are simultaneous.

The subsequent study of the repeated *mentor-mentee* interaction explores different combination of strategies. For the first example, we employ the grim strategy as a possible strategy for the *mentee* and the tit-for-tat strategy as a possible strategy for the *mentor*. The strengths of each strategy mentioned above are considered appropriate for the specific interaction. In addition, for the second example, we employ the tit-for-tat strategy for both the *mentor* and the *mentee* and provide a comparison analysis.

We also consider an additional strategy for the *mentor* to investigate options for cheating. Specifically, we consider a non-cooperative strategy referred to as *cheat-and-return* strategy, that the *mentor* employs against the grim trigger strategy and the tit-for-tat strategy (that are considered cooperative strategies) employed by the *mentee*. The cheat-and-return strategy gives the opportunity to the *mentor* to defect from cooperation and not provide the required quality. Since a *mentor* cannot in reality leave the *mentor-mentee* relationship (if the *mentee* selects to interact with the particular *mentor*), it returns to the interaction and accepts the *mentee's* punishment, if any. Example 3 will further explore the interaction of the two entities based on the particular strategies.

In summary, example 1 presents the numerical representation of the interaction of the *mentor* and *mentee*, who employ the tit-for-tat strategy and the grim trigger strategy respectively, example 2 presents the case that they both employ the tit-for-tat strategy and example 3 explores the *cheat-and-return* strategy by the *mentor* and appropriate responses by the mentee. Examples 1, 2 and 3 are numerically elaborated at the end of this section after discussing the use of the present value to quantify the payoffs in repeated interactions.

Since there is a finite number of *mentors* participating in the knowledge exchange platform, then it is expected that there will be repeated use of the same *mentors* for the same learning objective, in which case the rating can be used to characterise a *mentor's* past decisions on learning delivery. If the rating is available to all *mentees*, then repeated *mentee-mentor* interactions can be interactions with the same *mentor* for the same learning objective but not necessarily the same *mentee*. That implies that the *mentor* selection is completed by considering a universal rating of a *mentor*, but this is not the case we will explore in the *mentor-mentee* interaction. We would like to explore the interaction between a specific *mentor* and a specific *mentee*.

We will assume a continuous interaction with the same *mentee* in this model to demonstrate the case where a specific learning objective can require repeated interactions (e.g. different levels of knowledge in the same field of expertise). Nevertheless, we will use a rating **r** that represents a cumulative characterisation of the *mentor's* behaviour when interacting with any *mentee* for the specific learning objective. We assume that the designer enables the ratings of the relevant *mentors* to be available to and accessible by the *mentees*.

For the repeated interaction model, we will employ the game theoretic approach known as *repeated games*. The model of a repeated game is designed to examine the logic of long-term interaction. It captures the idea that a player will take into account the effect of the current decisions and their effect on the other player's future behaviour. Repeated game models aim to explain phenomena like cooperation, threats and punishment. The repeated game models offer insight into the structure of behaviour when individuals interact repeatedly, a structure that may be interpreted in terms of social norm. The results show that the social norm needed to sustain mutually desirable outcomes involves each player *threatening* to punish any other player whose behaviour is undesirable. Each player uses threats to warn the opponent that such punishment may follow if the threats are credible and if there is sufficient incentive for the player to carry out his threat.

Thus, punishment depends on how players value their future payoffs and it may be as harsh as lasting forever, or as mild as lasting for only one iteration of the interaction. We will explore these options in the examples at the end of this section.

As mentioned earlier in the chapter, the model of a repeated game has two kinds: the horizon may be finite, i.e. it is known in how many iterations the game ends, or infinite, i.e. the number of game iterations is unknown. The results in the two kinds of games are different. In the finite version of the *mentor-mentee* interaction, the interacting entities, and in particular, the *mentors*, will always tend to cheat in the last period of the game for the same reasons that cheating in the normal form game was the equilibrium choice. If they will cheat in the last iteration, then rational thinking implies that they would also cheat in the iteration before the last. The thinking actually holds all the way to the first iteration. So, in a finite horizon version of the game, the *mentors* would still not be motivated to cooperate, i.e. to deliver quality learning.

An infinite version of the *mentor-mentee* interaction plays out differently, because, there is no knowledge of the last period of interaction. As we will demonstrate later on, the numerical representation of an infinite horizon game model shows that there exists motivation for the *mentors* to cooperate. The main idea is that if the game is played repeatedly, then the mutually desirable cooperative outcome is stable because any deviation will end the cooperation, resulting in a subsequent loss for the deviating players that outweighs the payoff from the finite horizon game. Thus, when applying the model of a repeated game to a specific situation or problem, we must first determine whether a finite or infinite horizon is appropriate, based on the characteristics of the realistic situation. As we have justified already, the considered example must be considered within the category of infinite horizon repeated game models.

Whether the *mentor* cooperates or not, we can calculate the payoff based on the *present value (PV)* of the interaction since this is a repeated interaction and the payoff is based on all the periods of interaction up to the current one, plus the discounted value of all future interactions in the current period. The measure of the *PV* is very important to help us compare different strategies that are describing future actions, by looking at their value evaluated in the present.

PV is the sum that a player is willing to accept currently instead of waiting for the future payoff (i.e. accept a smaller payoff today that will be worth more in the future) similar to making an investment in the current period that will be increased by an interest rate in the next iteration. This is a popular method of evaluating a repetitive (possibly infinite or of unknown horizon) sequence of actions at a certain point in time.

Therefore, if the *mentor's* payoff in each period is expected t be equal to 1, then the *mentor* would receive 1 for the current period, but would receive, a value less than 1 to represent the payoff from the next period. The payoff for the next iteration that the *mentor* would be willing to accept in this period as the discounted payoff from the following period of interaction would be $(1 \cdot x)$, where $x \in [0, 1]$. Similarly, if the *mentor's* payoff in the next period is equal to P, then in this period, the discounted payoff would be $(P \cdot x)$, where $x \in [0, 1]$.

Following the same logic, if we have a payoff of 1 and that would make the payoff for the following interaction, payable in the current period equal to x, then the payoff in two periods would have to be discounted to x^2, the payoff in three periods would have to be discounted to x^3 and so on. Therefore, in the current period, the payoff the player, or in this case the *mentor*, would be willing to accept would be $1 + x + x^2 + x^3 + x^4 + \ldots$, until infinity.

For an infinitely repeated game, and considering the payoff of each interaction to correspond to the utility function UX_i for the *mentee* and UX_j for the *mentor*, then the present value in the current iteration is, for the *mentee*, equal to $(1 + x + x^2 + x^3 + x^4 + \ldots) \cdot UX_i$, or, for the *mentor*, equal to $(1 + x + x^2 + x^3 + x^4 + \ldots) \cdot UX_j$.

We can replace the parenthesis $(1 + x + x^2 + x^3 + x^4 + \ldots)$ with the sum of an infinite geometric series, which is equal to $\left(\dfrac{1}{1-x}\right)$. Thus, the present value in the current iteration is $\left(\dfrac{1}{1-x}\right) \cdot UX_i$ for the *mentee* and $\left(\dfrac{1}{1-x}\right) \cdot UX_j$ for the *mentor*, where x is often referred to as the discount factor. In case we will need to deal with the sum of a finite series, then consider this to be equal to $\left(\dfrac{UX_i \cdot (1-x^n)}{1-x}\right)$ for the *mentee* and to $\left(\dfrac{UX_j \cdot (1-x^n)}{1-x}\right)$ for the *mentor*, where n represents the number of iterations.

5.8.1   Three Numerical Examples of Repeated Interactions

In this section, we explore numerically three examples of the *mentor-mentee* interaction as this occurs repeatedly, when well-known repeated strategies are used.

The first example deals with the case when the *mentee* uses the grim trigger strategy and the *mentor* uses the tit-for-tat strategy for a repeated interaction game. We assume a history of cooperative moves in the past, and we compute the PV of both the *mentor* and the *mentee* in the present iteration. The *mentor* begins the move and plays cooperatively, which is the first move of a tit-for-tat strategy. Since the *mentee* plays the grim trigger strategy, which implies that the *mentee* will continue to interact with the *mentor*, unless it detects a cheating behaviour, neither of the entities will be motivated to deviate from cooperative behaviour. The *mentor* will continue mimicking the behaviour of the *mentee*, because of the tit-for-tat, and the *mentee* will continue to interact cooperatively. Hence, the two entities will continue cooperating forever. The present values represent the payoffs:

$$PV_{\text{mentor}} = \left(\frac{1}{1-x}\right) \cdot UX_j = \left(\frac{\lambda - \kappa}{1-x}\right)$$

$$PV_{\text{mentor}} = \left(\frac{1}{1-x}\right) \cdot UX_i = \left(\frac{r \cdot v - c}{1-x}\right)$$

We will refer to first strategy as conditional cooperation strategy. In this strategy, the *mentee* will terminate the interaction, if non-cooperative behaviour is perceived. That will cause the *mentor's* rating to be decreased for all subsequent interactions of the specific learning objective. The reason is that, because of the way in which the rating is calculated, the lack of positive rating results in the decrease of the overall rating. We will explore the ways in which the rating varies, in the following section. Note that since we consider a repeated interaction, e.g. online classes for the same short course, then each class is considered one interaction and will have its own option for a positive rating.

Example 2 considers that both the *mentor* and the *mentee* employ the tit-for-tat strategy. Again, we assume a history of cooperative moves in the past, and we compute the PV of both the *mentor* and the *mentee* in the present iteration. Since the strategy begins with a cooperative move from the *mentor* and the *mentee* mimics this because of the tit-for-tat strategy, the two entities will once more continue their cooperation forever. The payoffs will be the same as in the previous example.

Example 3 considers deviating from the previous cooperative types of behaviour by allowing in the current period a cheating move for the *mentor*. The *mentee* can employ either of the grim trigger strategy or the tit-for-tat strategy. Given the *cheat-and-return* strategy that we have presented earlier as an option for the *mentor*, let the *mentor* cheat in the current period and return to a cooperative behaviour in the next period. If the *mentee* employs the grim trigger strategy then, when the cheating behaviour is detected in the next period, the *mentee* will leave the interaction with the specific *mentor* forever. That present values are indicated below:

$$PV_{\text{mentor}} = \left(\lambda - \kappa'\right) + x \cdot \left(\frac{1}{1-x}\right) \cdot 0 = \left(\lambda - \kappa'\right) > \left(\lambda - \kappa\right)$$

$$PV_{\text{mentee}} = \left(r \cdot v' - c\right) + x \cdot \left(\frac{1}{1-x}\right) \cdot 0 = \left(r \cdot v' - c\right) < \left(r \cdot v' - c\right)$$

In the case that the *mentee* employs the tit-for-tat strategy, then when the cheating behaviour is detected in the next period, the *mentee* will leave the interaction but will return to the interaction after one period because it will detect a cooperative behaviour by the *mentor* again. According to the *cheat-and-return* strategy, the *mentor* will immediately return to cooperation after one period of cheating.

$$PV_{mentor} = (\lambda - \kappa') + x \cdot \left(\frac{1}{1-x}\right) \cdot (\lambda - \kappa) = (\lambda - \kappa') + x \cdot \left(\frac{\lambda - \kappa'}{1-x}\right)$$

$$PV_{mentee} = (r \cdot v' - c) + x \cdot \left(\frac{1}{1-x}\right) \cdot (r \cdot v - c) = (r \cdot v' - c) + x \cdot \left(\frac{r \cdot v - c}{1-x}\right)$$

The second *mentee* strategy of the third example can also be referred to as one-period punishment strategy since it only considers punishing the *mentor* for one period and then returning to continue cooperation. In terms of rating, the *mentee* will not give a positive rating for the specific interaction if non-cooperative behaviour is detected on behalf of the *mentor*. However, the *mentee* will still move on to the next interaction, i.e. to the next class and continue the learning, and given the *mentor's* cooperative behaviour in the next interaction, a positive rating is still an option for the *mentee* to give to the particular *mentor*.

These values show in general, that there is value in following a cooperative behaviour by both the interacting entities, when there is a repeated interaction of unknown horizon. If the *punishment for a cheating behaviour* is mild, i.e. one period of abstaining from interaction, then the motivation to cooperate is not as strong. On the contrary, there is strong motivation for the *mentor* to cooperate if the *punishment* by the *mentee* is stronger. Since we need to consider how the positive rating r affects this interaction as well, we will come back to the repeated interaction analysis with a different approach that considers r in the next section.

5.9   Adding a Rating Variable to Support Mentor Selection

The *mentor-mentee* interaction and the resulting payoff may be modelled such that it reflects a preference of the *mentee* towards *mentors* that do not often defect from cooperation during learning delivery. We employ the idea of allowing the *mentee* to enjoy an adaptive utility, i.e. a utility that can adapt to a specific changing variable. The specific variable is in fact the rating r of a *mentor*, such that the *mentee's* decision, regarding which *mentor* to select, considers a *mentor's* probability to defect from cooperation based on the acquired knowledge collected over the course of the repeated interactions of the *mentor* with *mentees* for the same learning objective. In this section, we will not explore the selection approach but we will explore the idea of introducing r, into the *mentee's* strategy of repeated interaction, such that the potential punishment for cheating behaviour is alignment with rating r.

In terms of design, the rating is based on an interface mechanism, which a *mentee* has the option of clicking on after an interaction. Clicking on the positive rating option gives a positive rating to the *mentor* for the specific delivery. This is quite similar to a *"Like"* or a *"Thumbs Up"* option, frequently seen on digital interfaces of social-type of applications through similar rating mechanisms. However, unlike most interfaces, the rating is calculated as a fraction from 0 to 1, where numbers closer to 1 show a higher probability of cooperation, whereas numbers closer to 0 show a higher probability to defect from cooperation.

The probability to defect from cooperation, or to cooperate, reflected through the rating mechanism, is considered as one of the parameters of the $UX_i(r)$ function, which a *mentee* uses to make the initial *mentor* selection. Thus, when the *mentee* must evaluate the predicted UX for selecting one or more of the available *mentors*, the evaluation should consider a probability of the *mentor* defecting from cooperation, i.e. cheating during delivery of learning. Next, we will explore options of including the rating more actively in the repeated interaction strategy.

The rating is calculated dynamically from observing past *mentor* behaviour reflected in the positive rating, which can be accordingly increased or decreased after each relevant interaction with a *mentee*. We will use an adaptive utility function to allow the *mentee to* make more informed strategic decisions during the interaction that considers the past.

This is achieved by using the variable r, representing the estimated probability of the *mentor* cooperating, such that $UX_i(r) = r \cdot v - c$, with c representing the probability of incurring unwanted costs, as evaluated subjectively by the *mentee*. This of course is a subjective measure but for such courses, the *mentors* are expected to deliver short courses with minimum additional resources needed. Nevertheless, the model includes this for completion purposes in order to show that the selection is a subjective decision by the *mentee*. Variable v represents the value the mentee gains from the interaction, where v represents the value from quality delivery, and v' represents the value from cheating *mentor* behaviour, such that $v' < v$.

To achieve this "behaviour monitoring", it is important that the designer allows the *mentee* to view the standing rating of a *mentor* and to be able to also rate a *mentor* after an interaction. The rating is only positive, i.e. the *mentee* cannot negatively rate a mentor. However, the rating history reflected in r may decrease as a result of the lack of positive ratings in time. Lack of positive rating after an interaction, which results in a lower overall r value does not necessarily imply negative delivery. This can simply be a result of *mentees* forgetting to rate positively. Furthermore, the designer cannot enforce a rating because we are looking for genuine positive ratings by the *mentees*. On the contrary, a high r value is definitely a representation of positive delivery, because *mentees* need to take the action of positive rating to allow the r value to increase. Also, the designer needs to be able to provide ways for saving and updating these metrics per *mentor* profile (the *mentors* will be notified of this rating as part of the initial participation agreement).

The evaluation of a *mentor* by each *mentee* (for the purposes of *mentor selection*) is different and subjective because of the c factor in the utility function. However, the r factor that is used in the *mentee's* utility function is objective, i.e. it is not decided by the *mentee* for the purposes of the selection process but given as an option by the digital interface designer.

The r factor is different for different *mentors* and different learning objectives. It represents a probability represented as a fraction between 0 and 1, and is updated if a positive rating is given by a *mentee* after an interactive period with the *mentor*, when delivery relevant to the specific learning objective has taken place. Therefore, the positive rating r updated by the platform's rating mechanisms, is based on a *mentor's* history of behaviour, and it is estimated as a *mentor's* probability not to defect from cooperation.

More specifically, the positive rating r is calculated as follows. In the case that the *mentor* receives a *click* on the positive rating interface feature, then:

$r = r_{\text{previous}} + r_{\text{previous}} \cdot UX_i$, if the resulting value of r is less than or equal to 1.

In the case the resulting value of r exceeds 1, then the $r = 1$. Note that UX_i is the utility of the *mentee* from interacting with *mentor* i, and the payoff value is subjective, but it is evaluated as a fraction between 0 and 1.

In the case that the *mentor* does not receive a *click* on the positive rating interface feature, then r is calculated as follows:

$r = r_{\text{previous}} \cdot UX_i$, which will result in a smaller fraction between 0 and 1 resulting from the multiplication of the rating with the utility function output.

By introducing the variable r, such that $r \in [0, 1]$, the *mentee* considers the *mentor's* history. Hence, the *mentee* adjusts any decisions related to their interaction, in an adaptive manner, i.e. by evaluating the expected utility based on a knowledge of the *mentor's* behaviour in past similar interactions, instead of basing the decision only on a subjective evaluation of the expected cost of the interaction.

Once the *mentee* makes the initial *mentor* selection decision, the *mentee* interacts with the selected *mentor* by specifying an appropriate strategy for this interaction. Since the interaction that we will consider is a repeated interaction, we will propose an adaptive strategy similar to the well-known repeated interaction strategies, which were discussed in the previous section. We will study how an adaptive approach towards the *mentor-mentee* interaction motivates a cooperative behaviour by the *mentor*, and how cheating behaviour affects the interaction when this adaptive approach is taken.

Considering the adaptive way in which the *mentee* takes a decision during the selection process, with the use of r, we propose a new adaptive strategy for the *mentee*. The adaptive strategy considers the repeated interaction approach of deciding each move by evaluating the *mentor's* behaviour in the previous period of interaction. The new adaptive strategy is not as strict as when the *mentee* employs the grim trigger strategy, but also not as lenient as the one-period punishment strategy employed where the *mentee* responds to a *cheat-and-return* strategy with a tit-for-tat approach.

Let the *mentee's* adaptive strategy be the following: cooperate as long as the *mentor* cooperates; if the *mentor* defects from cooperation, then do not give a positive rating for a y number of periods, where y is aligned with the positive rating r. After that, return and cooperate again (which would then imply a positive rating for the remaining number of interactions). Let the number y be equal to 1 if $r = 1$, or, to $\left\lceil \dfrac{1}{r} \right\rceil$ otherwise. The brackets symbolise rounding the fraction up to the next integer value, such that y is a whole number that can be translated into rounds that the *mentee* will apply the punishment of not interacting and consequently not rating the *mentor* positively.

This would result in a *mentor* with a lower overall r value, to suffer a lack of positive rating for more periods of interaction, whereas a *mentor* with a higher r value, to experience less number of periods of interaction with no positive rating (minimum punishment is 1 period, similar to the one-period punishment strategy). Let us refer to this strategy as the *adaptive-return* strategy employed by the *mentee*.

In the repeated *mentor-mentee* interaction game, the adaptive-return strategy for the *mentee* dictates that if the *mentor* defects from cooperation, the *mentee* punishes the *mentor* by withholding the positive rating for a y number of periods of interaction, before returning back to cooperation, and selects to rate the *mentor* positively again. The value of y is a *mentee*-generated value and is defined next:

$$y = \begin{cases} 1, \text{if } r = 1 \\ \left[\dfrac{1}{r}\right], \text{otherwise} \end{cases}$$

It is important that the *mentor* has a motivation to cooperate with the *mentee*, when the *mentee* employs the adaptive-return strategy. As we have seen earlier in the chapter, a strategy that employs a punishment that lasts forever is the strategy that provides the strongest motivation to cooperate (since the punishment is the biggest). We discuss next how when the *mentee* employs the adaptive-return strategy, the *mentor* is always at least as motivated to cooperate with the *mentee* as when the *mentee* employs the one-period punishment strategy, and that further, in most cases the *mentor* is more encouraged than these minimum motivation bounds provided when the one-period punishment profile is used.

Given a history of the game where both the *mentee* and the mentor have cooperated in the past, and the *mentee* employs the adaptive-return strategy, the *mentor* has two options in the current period: to either cooperate or to cheat in the quality of delivery of learning. The following section provides the quantification of this interaction by looking at the present value for the *mentor* and the *mentee* when the *adaptive-return* strategy is employed. The *mentor* is modelled both as a cooperative and a non-cooperative player in order to compare the resulting utility values.

5.9.1   Evaluating Adaptive Future Interactions

We consider the present value process in order to evaluate the payoff of the game for both the *mentor* and the *mentee*, when the *mentor* cooperates is as follows:

$$PV_{mentor} = \left(\frac{(\lambda - \kappa) \cdot \left(1 - x^{y+1}\right)}{1-x} \right) + x^{y+2} \cdot \left(\frac{\lambda - \kappa}{1-x} \right)$$

$$PV_{mentee} = \left(\frac{(rv - c) \cdot \left(1 - x^{y+1}\right)}{1-x} \right) + x^{y+2} \cdot \left(\frac{rv - c}{1-x} \right)$$

We have used the sum of the finite geometric progression to calculate the present values for the period corresponding to y, as defined previously, and the sum of the infinite geometric progression to calculate any subsequent behaviour expected to continue to the unknown horizon of interactions. Just as a reminder, the general form of the finite geometric progression formula is, for the *mentee*, $\left(\frac{UX_i \cdot \left(1 - x^n\right)}{1-x} \right)$, and for the *mentor*, $\left(\frac{UX_j \cdot \left(1 - x^n\right)}{1-x} \right)$, for a number n of iterations.

The sum of a finite geometric progression is used to calculate the discounted value for the first $y + 1$ periods of interaction, i.e. the current period and the subsequent y periods for which the punishment would hold in case of defecting from cooperation. Then, we add the remaining infinite sum for the remaining periods of interaction.

If the *mentor* cheats, on the other hand, then the payoffs based on the present values calculated for the mentor and the *mentee* are:

$$PV_{mentor} = \lambda - \kappa' + \left(\frac{0 \cdot \left(1 - x^y\right)}{1-x} \right) + x^{y+2} \cdot \left(\frac{\lambda - \kappa}{1-x} \right)$$

$$PV_{mentee} = r \cdot v' - c + \left(\frac{(0) \cdot \left(1 - x^y\right)}{1-x} \right) + x^{y+2} \cdot \left(\frac{r \cdot v - c}{1-x} \right)$$

Consider the case that y equals to 1, i.e. the minimum number of periods that can be imposed as punishment. The motivation for cooperation is not as strong as where the number of periods of punishment is higher. For only one period of punishment, the *mentor* may be able to afford a lack of positive rating without affect the overall rating. However, as the number of periods of punishment increase the effect on the overall rating increases with a negative effect on the overall reputation of the *mentor*. Since the punishment is aligned to the rating, there is strong motivation for the *mentor* to try to keep the overall rating high. When the rating is closer to 1, then, in case there is lack of positive rating, the negative effect is not such that it has a major effect on the overall reputation and limits the future interactions of the *mentor* with other *mentee*s as well.

The calculations for the present value allow us to compare the payoffs as quantified by the utility functions, for the various types of *mentor* behaviour over the repeated set of interactions. The *adaptive-return* strategy by the *mentee* allows for the flexibility in behaviour such that it passes the control of the *mentor-mentee* relationship to the *mentor*. The more that a *mentor* cooperates, the more the *mentee* will cooperate, the highest the rating will be and the less the number of periods of punishment, in case of cheating behaviour at any point of the interaction. Conversely, the more a *mentor* cheats, the less the positive rating by the *mentee* and consequently, the higher the number of punishment periods will be, which may result in negative effects across interactions of the *mentor* to other *mentee*s as well.

The adaptive-return strategy, where the rating r is used was part of the strategy, generates a range of punishments, which can be at least as harsh as the one-period-punishment strategy and not as harsh (or at most as harsh) as the punishment generated by the conditional cooperation strategy.

The rating mechanisms have provided an active, peer-to-peer monitoring that provides motivation to provide quality learning over such a mentoring, knowledge-exchange platform. The ease of navigation and learning is evident from the simplicity of the mechanism and the learning that will take place is expected to satisfy the *mentee*s' requirements. Thus, because of this mechanism, the user experience of the *mentee* and of the *mentor* will be positive in interacting through this digital interface. The measure of sociability will also be highly ranked because of the ease of usage of this mechanism and the positive emotions that will be evoked from *mentor-mentee* interactions with the use of the specific design feature.

5.10 Discussion

We have explored the interaction of two types of users across a knowledge exchange platform, where the two types of users interact over a digital interface. The peer-to-peer environment of the specific platform does not allow for a central authority to monitor and control the quality of knowledge exchange. Thus, a design feature that allows for positive rating is introduced in order to motivate the users of this digital interface to act cooperatively and deliver (or receive) learning based on the advertised quality. Given the requirements, the chapter outlined two models that make use of the positive rating such that the design motivates the required cooperation.

Primarily, the chapter presented a model of interaction between one *mentee* selecting between a number of *mentors* based on a simple rating scheme that easily adapts to the *mentors'* past mentoring success. The model examines the relationship between a *mentee* and a number of *mentors*, when all entities actively participate in an interactive mentoring platform and the *mentors* can satisfy a specific learning objective required by the *mentee*. The *mentee* is thus required to select a *mentor,* or a group of *mentors*, based on the learning requirements. The model is based on the evaluation of a utility function by the *mentee,* which reflects the subjective evaluation of the *mentors* that want to participate in the interaction.

The second model describes the use of the rating mechanism to guide the interaction between the selected *mentor* and the *mentee* in a *mentor-mentee* interaction, given that there exist incentives for a mentor to cheat on the quality promised for the delivery of learning. The chapter takes a closer look at the interaction between the *mentee* and the selected *mentor,* and provide more details related to the use and effect of the positive rating feature, introduced by the design.

In terms of the mathematical analysis, the interaction has been explored as a one-shot game theoretic model, which led to the conclusion that cooperation between the mentor and the mentee is not supported in terms of rational decision-making, because the mentor always has an incentive to cheat if there are no consequences of such an action in subsequent interactions.

However, since in realistic scenarios previous outcomes of the interaction affect the behaviour of the two entities in a present interaction, we extended the one-shot model to a repeated interactions model. The repeated game model was introduced to investigate whether cooperation could be reached under different conditions, i.e. when actions from the mentor in one period of interaction, could have consequences on the future interactions. In particular, we present a repeated game model that captures the case where players have a window of the previous history of the game, which affects the current player actions. Thus, we investigate how the behaviour of both the mentor and the *mentee* changes if we consider a repeated interaction between the two entities, where an infinite horizon is considered. An adaptive strategy based on the design feature of allowing the *mentee* to positively rate the mentors results in motivation for a cooperative behaviour.

The purpose of both models is to present how a satisfying user experience can be achieved through this simple mechanism, i.e. positive rating mechanism, by breaking down the decisions taken during a simple selection of the preferred *mentor* or *mentors* by a *mentee* for a specific learning objective, and the subsequent interaction during the delivery of the learning objective. When we refer to the mechanism as simple, we mean simple in terms of design as well as simple in terms of updating the necessary computations. Nevertheless, the value of having this kind of motivation to guide interaction is high.

In conclusion, we would like to once more highlight the positive effect that a simple design decision, such as to include positive rating as part of the design of a knowledge exchange platform, has on the user experience. This is particularly true when viewed as an attempt to improve the sociability element of the interface. The specific element improves the emotional response of the interface users in terms of their behavioural reactions. A user that has the role of the *mentee* will experience easier navigation in terms of selection of mentors to interact with and confidence that they will be given enough information to make positive decisions for their learning. A user that has the role of the mentor is aware of the rating system and can better organise the delivery of the learning objectives to achieve positive ratings, making the use of the platform easier and comfortable.

References

1. Antoniou J, Pitsillide A (2006) Radio access network selection scheme in next generation heterogeneous environments. In Proceedings of the IST Mobile Summit, 2006
2. Antoniou J, Papadopoulou V, Pitsillides A (2008) A Game Theoretic approach for network selection, Univ. of Cyprus, Tech. Rep. TR-08-5
3. Norman DA (2004) Emotional design: why we love (or hate) everyday things. Basic Books, Cambridge, MA. ISBN 0-465-05135-9
4. Lowgren J, Reimer B (2013) The computer is a medium not a tool: collaborative media interaction design. Challenges 2013(4):86–102
5. UNICEF (2015) Learning and knowledge exchange, [Online] https://www.unicef.org/knowledge-exchange/
6. Urrutia ML, White S, Dickens K, White S (2015) Mentoring at scale: MOOC mentor interventions towards a connected learning community, EMOOCs 2015 European MOOC Stakeholders Summit, Belgium, 18–20 May 2015
7. Dabner N (2011) Design to Support distance teacher education communities: A case study of student-student e-mentoring initiative. Society for Information Technology & Teacher Education International Conference, Mar 07, 2011 in Nashville, Tennessee, USA ISBN 978-1-880094-84-6 Publisher: Association for the Advancement of Computing in Education (AACE)
8. Gintis H (2000) Game theory evolving: a problem centered introduction to modelling strategic interaction. Princeton University Press. 0-691-00942-2

Chapter 6
Dealing with Emerging AI Technologies: Teaching and Learning Ethics for AI

6.1 Introduction

The chapter deals with the teaching and learning aspects of ethics for emerging new technologies such as artificial intelligence (AI). It is important to recognise that there are ethical issues associated with the development and use of such technologies. Furthermore, as we are moving into an era where the development and use of AI is expected to increase across many application areas, we need to consider ethics of developing and using AI as part of necessary learning and teaching practices. AI is not a stand-alone technology but a technology that is integrated with society and people, and needs to be understood in such an inter-disciplinary manner.

Integrating ethics into a practical study of technology and computing is challenging in its own right, and to do that, we need to return back to basics of learning and try to break down the study of ethics, such that they can be understood, applied and analysed step by step.

Teaching and learning ethics is a challenging task [16]. Several issues arise, especially when we aim to focus on teaching and learning ethics for a particular field of study, AI in this case. It is important to investigate whether there is a need for ensuring that the learners have covered any prerequisite knowledge, if we decide that there are prerequisites. On the other hand, if we decide to approach this learning as an autonomous and independent piece of learning with no prerequisites, then we need to investigate ways in which we can approach the learning objectives from scratch, while ensuring that by the end of this task, learning will have successfully occurred.

The chapter will approach this type of learning as independent, i.e. without any prerequisite knowledge being assumed or required. However, we will make the assumption that ethics for AI is an area of learning that we expect to be interesting for professionals and scientists, that already have a level of understanding of the concepts we will cover, through previous studies and experiences. In order to break

© Springer Nature Switzerland AG 2021
J. Antoniou, *Quality of Experience and Learning in Information Systems*,
EAI/Springer Innovations in Communication and Computing,
https://doi.org/10.1007/978-3-030-52559-0_6

down the study of ethics such that they can be understood, applied and analysed, the chapter will try to employ Bloom's taxonomy of learning as a useful learning approach that can be applied to the area of ethics for AI.

Bloom's taxonomy of learning [4] is often viewed as a series of educational objectives that has been used to develop learning objectives in multiple dimensions including learning ethics [8] and improving critical thinking and decision-making [11]. Although, there are debates on whether and in which way ethics can be taught, there exists knowledge around ethics that needs to be disseminated across the AI development and usage areas of research and innovation. With the increase in usage of emerging technologies, including AI, the integration of technology with society and everyday life dictates that the human element is considered in greater focus, and ethics as part of the development and usage of such technologies is an effective way to approach this issue.

In this chapter, we will present an approach that will attempt to understand how ethics relates to the theory and practice of AI-related development and use, in an attempt to identify what needs to be learned in a future AI-enhanced society.

6.2   Learning Steps and Objectives

Although there are concerns regarding the ethics that need to be considered for AI development and use, there does not seem to be an agreement on what AI developers and users need to know in practice, in terms of ethics. The urgency in identifying the learning steps for AI ethics comes especially from the fact that there seems to be a consensus that we are moving forward into a future where the presence of these technologies is expected to be even more prominent [12].

The chapter explores an approach to learning AI ethics such that the learning process is aligned with a set of learning steps derived from Bloom's taxonomy, a taxonomy often used to develop learning objectives [4]. Specifically the steps consider the following sequence of actions, *remembering*, *understanding*, *applying*, *analysing*, *evaluating* and *creating*. The idea is that by moving through the actions, one step at a time, learning can be achieved. Each step can be broken down into specific content-related activities aligned with the overall effective learning of the bigger concepts; ethics in this case. Given that the context of the ethics is technology, activities and objectives can be chosen such that the context is made use of.

We can use the steps of Bloom's taxonomy of learning as a basis to design learning objectives such that they are aligned with the taxonomy, while at the same time considering findings from literature that explores the ethics of the application of AI research and innovation. The taxonomy can further help us characterise ethical categories or clusters that may be used as the basis of investigating the knowledge, understanding and use of ethics by AI practitioners. Going through this exercise is important in order for us to be able to identify what needs to be learned and how learning and teaching can go about achieving this. Moreover, as a future step, the characterisation of the learning of AI ethics into learning objectives may lead to meaningful assessment and further educational planning.

The chapter will investigate the creation of targeted learning objectives for ethics in AI, while allowing for the development of an assessment of the learning. This is a challenging task, and one of the challenges of deciding on the core ethical knowledge that AI practitioners should know is the lack of clarity in what the development and use of such systems entails. In addition, it is unclear what should in turn be known in order to best develop and use such systems in the future. It is, therefore, important that appropriate ethical competence is considered and ensured.

In a recent publication that discusses the significance for ethics in technology [14], the authors investigate how Smart Information Systems based on AI and Big Data Technologies can be used to support sustainable development while at the same time raise some ethical concerns, which are listed as important ethics to consider in relation to the specific technologies. In alignment with the aforementioned publication, the chapter will reproduce the listed ethical concerns as examples of ethics that need to be learned by AI practitioners. Specifically, the list of ethics to consider include: *equality*, with a consideration of inequalities that may be heightened due to the use of AI systems; environmental *sustainability* concerns; *privacy* and *confidentiality*; *discrimination* and *bias*; *responsibility*, *transparency*, data *accuracy*, and *trust*; *security* issues and *availability* of resources.

Looking at Bloom's taxonomy of learning, the first two levels are about *remembering*, i.e. what may be considered basic or innate knowledge, and *understanding*, i.e. a process of enhancing existing knowledge. Both levels are evident in Bloom's original taxonomy and the revised one [1]. Remembering deals with being able to identify, to name, to memorise and to label and for such tasks, the concepts to be learned need to be clear, factual and not open-ended but resolved in some way. Moving on to understanding, the learning can proceed into describing these facts, classifying them and even explaining them.

Considering ethics as a list of facts, we need to refer back to, or *remember*, the list of human freedoms [7]. The use of AI technologies may allow application users' behaviour to be put under close scrutiny and surveillance, leading to infringements on privacy, freedom, autonomy and self-determination [18]. There is also the possibility that the increased use of algorithmic methods for societal decision-making may create a type of technocratic governance [6]. In general, the list of human freedoms includes such concepts as: the right to a private life, the right not to be discriminated against, the right to education, etc. Furthermore, we need to further *understand* that in order to ensure these human freedoms, the development and use of AI technologies must be responsible (on behalf of the developer) such that there exists a sense of trust (on behalf of the user). Therefore, the first learning objective could be: *to be able to recognise and understand that AI development and use must be ethical.*

Bloom's taxonomy continues to the third stage, that of *application*. This is more about being able to demonstrate learning, to put it into practice or be able to use it in context. This learning objective is about developing ethical judgement and moving from simple understanding to becoming aware of the ethical dimensions of decision-making within specific situation. A more practical aspect of learning ethics may include the use of practical scenarios, where ethics are used in context and the learner can demonstrate decision-making. For example, the scenarios can focus on idea of control of data [13], which includes decisions about privacy, consent and security that must be made.

In fact, AI application developers have a responsibility to the application users to ensure compliance, accountability and transparency of their AI applications [10]. However, when the source of a problem is difficult to trace, owing to issues of opacity, it becomes challenging to identify who is responsible for the decisions made by the AI. The concern surrounding privacy can be put down to a combination of a general level of awareness of privacy issues and the recently introduced General Data Protection Regulation (GDPR). Closely aligned with privacy issues are those relating to transparency of processes dealing with data. Therefore, the second learning objective could be: *to be able to make informed decisions about ethical concerns in AI development and use scenarios.*

Bloom follows the *application* phase with that of *analysis* to show the additional requirements beyond simply developing the sensitivity to recognise an ethical decision but needing to dive into more specific knowledge in order to inform the decision. Analysing is about criticising as part of learning, through careful appraisal, comparison and examination. Analysing activities should provide the opportunities to question and test concepts prior to reaching a conclusion. Ethics for AI development and use, as a new area of study, offers a number of opportunities to question concepts. Consider for example the situation where the identification of the level of reliability of a specific AI application is needed, or of a specific dataset that may be used by an AI application. Examining this problem requires the learner to look into the accuracy of data and the accuracy of the algorithm used, before coming to a conclusion [17].

Reliability is further linked to the requirements of diversity, fairness and social impact because it addresses freedom from bias from a technical point of view. The concept of reliability, when it comes to AI, refers to the capability to verify the stability or consistency of a set of results [9]. The importance of reliability lies in the fact that once data is shown to be reliable, then it can be shown to be valid – validity of AI is an essential prerequisite to the use of the specific technology [5]. If a technology is unreliable, error-prone and unfit-for-purpose, adverse ethical issues may result from decisions made by the technology. The accuracy of recommendations made by AI applications is a direct consequence of the degree of reliability of the technology [3]. Therefore, the third learning objective could be: *to be able to critically discuss and analyse potential ethical issues related to AI development and use examples.*

The last two levels of Bloom's taxonomy are *synthesis* and *evaluation*. These appear as *evaluation* and *creation* in the revised version of the taxonomy. In either case, we view these two stages as a combined requirement in our categorisation process. We interpret the last two stages as aligned with the general objective of aiming to maintain an ongoing and sustainable commitment towards ethical development and use of AI technologies.

Specifically, the idea of *evaluating* encompasses activities related to assessing, arguing and predicting aspects of the AI ethics related to an application, whereas, *creating* moves further into activities such as design, constructing and proposing solutions for AI application that can ensure the required ethics. Some example ethics that we can consider to demonstrate this objective include bias and power asymmetries. Specifically, we can consider how an AI application may be *assessed* in terms of bias in order to avoid discrimination or power asymmetries during its use, and further, how we may propose a design that ensures such ethical use.

Reaching this level of learning in relation to ethics will provide confidence for AI professionals that they are able to maintain a sustainable ethical approach towards the development and use of AI moving forward into the future. An additional ethical concern is that biases and discrimination can contribute to inequality. Data used to train algorithms may exclude some minorities who do not have access to the Internet or social groups excluded from society. In this way, the analyses carried out by the use of algorithms may not be representative of the whole population under examination [15]. Some groups that are already disadvantaged may face worse inequalities, especially if those belonging to historically marginalised groups have less access and representation [2]. Thus, it is important to allow creating as part of the learning of AI ethics to ensure not only the consideration of such situations but also the design of ethical AI applications and avoid such inequalities. Therefore, the fourth learning objective could be: *to be able to assess potential ethical issues that may arise from specific AI development and use and to be able to propose alternative ethical designs.*

Therefore, the resulting list of learning objectives to be considered for teaching and learning ethics for AI development and use consists of four learning objectives, as follows:

1. *A student learning ethics for AI development and use should be able to recognise and understand that AI development and use must be ethical.*
2. *A student learning ethics for AI development and use should be able to make informed decisions about ethical concerns in AI development and use scenarios.*
3. *A student learning ethics for AI development and use should be able to critically discuss and analyse potential ethical issues related to AI development and use examples.*
4. *A student learning ethics for AI development and use should be able to assess potential ethical issues that may arise from specific AI development and use and to be able to propose alternative ethical designs.*

Thus, teaching and learning ethics is essential for AI developers and users. Traditionally, however, software development education focuses on technical aspects and does not consider the element of ethics as a major component (even though ethical and legal issues may be briefly mentioned in professional courses). The four learning objectives discussed above should be incorporated in software development related courses, especially where technologies such as AI are concerned. The reason is that emerging technologies such as AI have the potential to impact society significantly, because they can affect a high number of citizens. AI products often use learning mechanisms, where software models are used to artificially interpret data and learn how to make decisions. In order to be able to learn from data collected, it needs to be a very large amount of data, in particular user data, and hence the potential high impact of AI on society.

6.3   Introducing Ethics in Technology Design

Incorporating ethics in technology design is a desirable strategy. As users come to understand the increasing impact of AI products in their lives, the knowledge that ethics were considered in designing and developing these products can put their worries somewhat at ease. A positive expectation created from this knowledge can improve the user experience while using such products. The challenge for most application designers is the cost of considering ethics as part of their design as this would add to the overall development cost. However, looking at this long term would benefit the developers.

Given that specific user audiences have a requirement for ethical design, it may be necessary to undergo training so that a developer can have a certificate for ethical design to convince its audience.

Consider the case of different AI developers. The AI developers make profit by developing, and making available for use, new AI applications, which are based on ethics-aware design. The developers must design the application by considering the ethics of AI, given the additional cost of considering ethics-aware design practices during their development methodologies, because of the user demand for ethical AI software.

Suppose the expected revenue from making the new AI application available for use is v for a specific user audience. Let the cost of developing the application and considering ethics as part of the design be c_1 per application if each developer approaches the design process without external training. This is often the case if such certification is not required by the specific user audience, since formal certification may incur additional cost. The additional cost, however, is a once-off cost, so for subsequent interactions, development costs would be free of this cost. If the developers undertake external training in order to receive the appropriate certification, the resulting cost of development in subsequent development instances will actually decrease given the training, and will be c_2 per application, where $c_2 < c_1$. Many developers are hesitant to consider external training if not required by the user audience because of the initial training costs.

Note the significance of the learning objectives, derived in the beginning of the chapter, for achieving the formal training suggested for enhancing the AI development skills.

The particular model needs to be approached as a long-term model, i.e. where each developer is planning to continue to develop AI applications, and the users that will continue purchasing the applications expect them to be ethically aware.

So, there are two types of costs in the ongoing development of applications, c_1 and c_2, where $c_1 > c_2$. Let us quantify these costs as compared to the profit from the development of the AI application. We will make the assumption that $\frac{v}{2} > c_1$ for this example scenario, i.e. the greater cost of designing for ethics without previous external training is less than half the expected revenue from selling the AI application. This may be considered as true given that no additional cost other than the implementation effort is needed for making available the AI software, but the revenue is made repeatedly as more and more users decide to purchase and use it.

Therefore, a rational developer would decide that since $c_2 < c_1$, then, if the user audience requires an ethical design, they would prefer to invest in external training and receive the corresponding certificate. This would support not only the specific development but also future development of new AI applications. There are also the developers that will not be required to get external training but will decide to consider ethics in-house and develop ethically aware applications with a recurring cost c_1 per application.

6.4   Introducing Unethical Behaviour in the Scenario

We should also consider developers who will resort to *unethical* practices to convince the specific customers' audience that they are ethical in order to avoid any additional cost. Such behaviour could take place in the following manner: the developers will try to mimic some ethical guidelines without really putting the effort to either take the training or do the in-house work for ethics alignment. However, they would only be able to do this if the user audience does not require certifications of ethical design, and if they find at least one ethical developer to mimic. Otherwise, the user audience will reject them since ethics-alignment is a requirement overall for this model (even if a formal certification is not required).

Consider the interaction of *two* (in order to study the simpler competition game between only two entities) developers producing a specific AI application for the same user audience (i.e. the AI applications serve the same purpose and have a very similar functionality). Consequently, the developers may find themselves in the following various scenarios:

The first scenario is that the user audience requires an ethics design certifications by the developers. Not attending external training for any one of the competing developers will result in zero (0) payoffs, since the requirement of formal certification is needed by the audience in order for the purchase of the AI product to go through. Thus, both developers decide to undertake the cost of external training. However, this results in subsequent development costs to decrease to c_2. In this scenario, the revenue is $\frac{v}{2}$ since two developers are now competing for the same user audience, and we will assume that there are no other factors affecting the users' decision. Therefore, they split in the middle in terms of which application to purchase. The payoff of each developer in this scenario is equal to their profit, i.e. $revenue - cost = \frac{v}{2} - c_2$, for each developer.

The second scenario is that the user audience does not require an ethics design certification. In this case, there is room for unethical behaviour to take place. Let us assume that the developers can decide whether to act ethically or unethically. In particular, the first developer may decide to avoid training costs and develop an original ethics design in-house resulting in cost c_1. On the other hand, the second developer may decide to mimic the first developer, incurring no costs by copying the ethical design. This *unethical* move by the second developer can eliminate any ethics design-related costs. Considering that other things equal, the second developer can end up receiving a revenue of $\frac{v}{2}$. In this scenario, the first developer will end up receiving a revenue of $\frac{v}{2}$, but there exists also cost of c_1, which results in a final profit, $revenue - cost = \frac{v}{2} - c_1$.

The third scenario considers the situation where the user audience does not require any certification, and both the developers want to act unethically in order to avoid incurring any additional costs. In the previous scenario, we have explored the case that one of them acts unethically and the other one acts ethically. In this scenario, we take a look at the case when neither of the developers takes any steps to produce an ethically aligned application as each is waiting to mimic the other one's design. The result is that neither of the developers manages to design an ethically aware application and are rejected by the user audience, resulting in zero (o) profit for both (no revenue and no cost).

Consider one round of the above interactions as described in the three examples (same options for rows and columns, where one developer is represented by the rows, and the other by the columns; the payoffs are represented as (row player, column player). In particular, Table 6.1 considers the options of each developer, both when there is a requirement for a certification or not. For example, the action *Ethics aligned designed for AI* considers both costs c1 and c2 to cover both cases. The requirement for certification only results in a competition, when both entities decide on the *Ethics aligned design for AI* move. For audiences that do not require certifications, all other combination of actions are possible.

6.5   A First Glance at the Competition Model

It is easy to analyse the above payoffs, by considering that $\frac{v}{2} > c_1 > c_2$. Thus, if a player is playing against a non-cheating player, they are inclined to cheat in a situation that ethics aligned design is not a requirement because that will result more profit. Similarly, if the developer believes that they are playing against a cheating

Table 6.1 Competing developers choosing between ethical and unethical moves

	Ethics aligned design for AI	Cheat (mimic opponent)
Ethics aligned design for AI	$\left(\dfrac{v}{2}-c_2,\dfrac{v}{2}-c_2\right)$ or $\left(\dfrac{v}{2}-c_1,\dfrac{v}{2}-c_1\right)$	$\left(\dfrac{v}{2}-c_1,\dfrac{v}{2}\right)$
Cheat (mimic opponent) $\left(\dfrac{v}{2},\dfrac{v}{2}-c_1\right)$		$(0,0)$

player, the only choice is not to cheat in order to avoid the zero payoff even though the other player will profit more. At a first glance, it appears that the combination of a *Cheat* decision with a decision for *Ethics design for AI* gives an equilibrium.

Note that what is represented here as the *Cheat* decision models any work that does not consider ethics as part of the development and use of AI, since this represents older approaches or even, in some cases, current approaches to development of AI applications. Research literature and ongoing project works show us that future development of AI needs to consider such ethical guidelines, either through company initiatives or through policy and regulation. Policy and regulation would impose a requirement on developers, and rational decision-making would naturally lead to a combination of decisions for ethics aligned design.

Returning to the analysis of the above model, it is important to consider the evolutionary stability of these options as they are repeated through time, since the developers are expecting to be able to continue developing AI applications for audience with ethics requirements, beyond the first development round. This changes the equilibrium and highlights better decisions for the developers.

6.6 Evolutionary Stability of the Competition Model

Next, we will generalise the payoff matrix, so that we can deduce whether it demonstrates elements of evolutionary stability. Consider Table 6.2 as the generalisation of the model's payoff matrix.

We observe that $a \neq c$, $b \neq d$, $a < c$ and $d < b$. This means that the ethical strategy does not strictly dominate the cheating strategy to show that there is evolutionary stability. However, given two rational players playing simultaneously, cheating may end up with payoff d rather than the required c for both players, so they will choose ethical behaviour to ensure that the avoid zero payoffs and any of the other payoffs a, b or c is better than d.

Table 6.2 Generalised payoffs of the competition model

	Ethics aligned design for AI	Cheat (mimic opponent)
Ethics aligned design for AI	$\left(\dfrac{v}{2}-c_2,\dfrac{v}{2}-c_2=(a,a)\right)$ or $\left(\dfrac{v}{2}-c_1,\dfrac{v}{2}-c_1=(b,b)\right)$	$\left(\dfrac{v}{2}-c_1,\dfrac{v}{2}=(b,c)\right)$
Cheat (mimic opponent)	$\left(\dfrac{v}{2},\dfrac{v}{2}-c_1=(c,b)\right)$	$(0,0)=(d,d)$

We will attempt to analyse this model over time. Let us assume that there is a probability b that there is a requirement for ethics design certification and both developers will take the trainings and decide on *Ethics aligned design for AI* with development cost c_2. Thus, $P - b$ represents the probability that there is no such requirement and unethical behaviour is possible, which is more interesting in order to investigate the developers' motivation for ethical behaviour without there being a requirement for it.

Let us also assume this probability b of undertaking training is broken down as the product of two other probabilities, such that $b = d \cdot g$. In this representation, d is the probability that the first developer will need to get training to fulfil this requirement and g is the probability that the second developer will need to get training to fulfil this requirement. Therefore, we have:

$$P - b = P - g \cdot d = g \cdot \left[d \cdot \left(\frac{v}{2}-c_2\right)+1-d \cdot \left(\frac{v}{2}-c_1\right) \right] + (1-g) \cdot d \cdot \left(\frac{v}{2}\right)$$

$$= \frac{v}{2} \cdot \left[g + (1-g) \cdot d \right] + g \cdot d \cdot (c_1 - c_2) - g \cdot c_1$$

Looking at the first line, the probabilities of both the developers are broken down into the probability of acting ethical or not, based on the payoff matrix. The second line is the simplification of this payoff probability expression. The interesting observation from the simplification is that the payoff probabilities seem to be favoured in the cases that either one of them cheats or both of them follow an ethics design strategy, which is true. The higher the probability for a requirement of ethical design, the more favoured the cooperative behaviour is, such that both developers decide on an ethics aligned design.

More specifically, consider the below expression, where a developer plays *ethically*, d fraction of the time with training and $(1-d)$ fraction of the time, the developer plays *ethically* but with in-house design:

$$d \cdot \left(\frac{v}{2} - c_2 \right) + (1-d)\left(\frac{v}{2} - c_1 \right) = \frac{v}{2} - \left[d \cdot c_2 + (1-d) \cdot c_1 \right]$$

Note that we consider probability d to be equal to $0 < d < 1$.

Now given, the first developer plays as above, developer 2 will also play *ethically* according to the first developer's moves, so that the probability g of playing *ethically* for the same user audience depends on the first developer's play such that the second developer will either play:

$$g \cdot \left[d\left(\frac{v}{2} - c_2 \right) + (1-d \cdot \left(\frac{v}{2} - c_1 \right)) \right]$$

or, in the case that there is no requirement:

$$(1-g) \cdot d \cdot \left(\frac{v}{2} \right)$$

As long as there is a probability to cheat, the evolutionary stability will favour the one-shot set of payoffs in Table 6.1, where the developer will cheat if there is sufficiently high probability not to end up with zero payoff.

If a developer is risk-averse, i.e. does not want to risk having zero payoffs, then, acting ethically is the right move over time. Moreover, if we consider that over time the choice of undergoing the training will result in higher payoffs than not having gone through it, since $c_1 > c_2$ in any new development opportunity, then a risk-averse player will undergo the training and select ethical behaviour in each competitive new development opportunity.

On the contrary, a risky player will hope to come across such risk-averse players, so that cheating behaviour will bring the highest reward. Over the course of time, there needs to be less opportunities for players to cheat in order for evolutionary stability to favour an ethical behaviour from all developers of AI, which will serve as a reassurance to AI application users, who will end up with a more satisfying user experience.

6.7   Motivation for Ethical Decision-Making

The simplistic analysis of two interacting entities has produced a motivation for a cooperative behaviour in which ethics play a significant role. As such, increasing the probability that lack of ethics consideration in development will result in less payoffs is important to push developers over time to demonstrate ethical decision-making behaviour. For example, motivation for two cooperating developers who want to behave ethically could be that by doing so they can receive funding for allowing the ethics committee to review their design. This will provide reputation value to the developers and a motivation for consumers to purchase their products. The consumers are not considered in this model, i.e. their preference is not factored into the motivation for the developers to make their design decisions, even though it will play a significant role for developers if it has an impact on the overall profit. Obviously, the assurance of ethical behaviour is attractive to potential users of AI products.

Suppose that we begin with the current state of AI software developers and assume that most of them do not practice ethical design. Considering our model, this would imply that most of the developers would choose the cheating behaviour, except for the risk-averse developers. Let's assume that a developer that is *ethical* will keep choosing "*ethical* behaviour" and a developer who is *unethical* will keep choosing *unethical* behaviour.

Let us consider that at this time $(1-e)$ of the developers are *unethical*, whereas a small fraction of the developers population, e, represents *ethical* developers.

So *unethical* developers will have a $(1-e)$ probability of finding across them an *unethical* developer and receiving a payoff of 0 and a probability e of finding an ethical developer and receiving a payoff of $\frac{v}{2}$, such that:

$$(1-e)\cdot(0)+e\cdot\left(\frac{v}{2}\right)$$

On the other hand, *ethical* developers will be paired with *ethical* developers with a probability e, and receive $\frac{v}{2}-c_2$ payoff, but will be paired against *unethical* developers with a probability $(1-e)$, and for those cases, they will receive $\left(\frac{v}{2}-c_1\right)$ payoff, such that:

$$(1-e)\cdot\left(\frac{v}{2}-c_1\right)+e\cdot\left(\frac{v}{2}-c_2\right)$$

Over time, if more developers decide to convert to ethical behaviour to avoid zero (0) payoffs, we will end up with:

$$e\cdot\left(\frac{v}{2}\right)<(1-e)\left(\frac{v}{2}-c_1\right)+e\cdot\left(\frac{v}{2}-c_2\right)$$

In fact, as the ethical developers increase in numbers, the cooperating fraction of the population, i.e. the ethical developers, will be closer to $(1-e)$ and the non-cooperating will be closer to e. Therefore, replacing the new fractions into the previous expressions, we get a future model of the population of developers, such that:

$$(1-e)\left(\frac{v}{2}\right)<(1-e)\left(\frac{v}{2}-c_2\right)+e\left(\frac{v}{2}-c_1\right),$$

which now depends on the cost c_2 to keep the equilibrium.

Ethical behaviour can only be motivated if $c_2 < \left(\dfrac{v}{2} - c_1 \right)$, which requires regulators to offer necessary incentives, e.g. funding, or new requirements, to support this evolutionary model in order to motivate the population towards ethical behaviour than not.

The need for ethical designers is evident in a world where users demand such behaviours since unethical designers cannot survive without being able to interact with ethical designers. The need for ethical designers, inspired by a user centric design environment, where ethical behaviour satisfies more users, is motivating for more and more professionals to act ethically. Therefore, in time, the $(1-e)$ population percentage is expected to increase and so will the overall payoff of ethical designers over time. Note that $\dfrac{v}{2} > c_1 > c_2$ results in non-zero payoffs for unethical designers even if their overall fraction decreases.

6.8 Discussion

User experience is often affected by the users' expectations of how their interaction with a service is expected to be, even before actually interacting with it. This is true, especially when such expectations are not so positive because of worrying about interacting with a new and unknown technology. Many users would practice caution when engaging with AI software products, simply because of the buzz that the technology carries. In the chapter, we have made the assumption that if policy allows for ethically aligned certifications to be given out for AI software designers, then most of the user audiences would require that certification. As relevant policies are put in place, most of the developers will have the necessary training done in order to receive a certification, or will be happy to receive external audit in order to convince their customers of their ethical development practices.

The chapter considers as a fact that there will always be some developers who will resort to unethical practices. The motivation of such behaviour is in order to convince their customers to trust them by offering an ethical façade without really engaging in ethical design practices and especially without incurring the relevant costs. Such behaviour is modelled as a decision of those developers to try and mimic some ethical guidelines without really putting the effort to either take the training or do the in-house work for ethics alignment. However, if they don't have the opportunity or they don't succeed in mimicking ethical behaviour, the user audience will reject them, and they will end up with zero gains from such actions. The chapter explores the idea of probability of finding ethical and unethical developers in a finite population of AI developers, and how there exist motivation across the population to increase the number of ethical developers as time progresses, according to evolutionary stability principles.

The findings of exploring a simple interaction pose new requirements on new approaches for teaching and learning ethics, which is becoming essential for AI developers (and users). Since this is a new trend, not supported by traditional models for the education of software developers, we propose to incorporate new learning objectives in such educational approaches to complement the technical aspects of the learning. The four learning objectives discussed in the chapter could be incorporated in software development related courses, especially where technologies such as AI are concerned. Emerging technologies such as AI have the potential to impact society and citizens significantly, and learning at the stage of development can affect user experience at the stage of usage.

References

1. Anderson LW (ed), Krathwohl DR (ed), Airasian PW, Cruikshank KA, Mayer RE, Pintrich PR, Raths J, Wittrock MC (2001) A taxonomy for learning, teaching, and assessing: a revision of Bloom's taxonomy of educational objectives (complete edition). Longman, New York
2. Barocas S, Selbst AD (2016) Big data's disparate impact. Calif Law Rev 104(671):671–732. https://doi.org/10.15779/Z38BG31
3. Barolli L, Takizawa M, Xhafa F, Enokido T (eds) (2019) Web, artificial intelligence and network applications. Proceedings of the Workshops of the 33rd International Conference on Advanced Information Networking and Applications, Springer
4. Bloom BS (1956) Taxonomy of educational objectives, handbook: the cognitive domain. David McKay, New York
5. Bush T (2012) Authenticity in research: reliability, validity and triangulation. Chapter 6 in edited "Research Methods in Educational Leadership and Management", SAGE Publications
6. Couldry N, Powell A (2014) Big data from the bottom up. Big Data Soc 1(2):205395171453927. https://doi.org/10.1177/2053951714539277
7. Council of Europe (1950) European convention of human rights and fundamental freedoms. European Court of Human Rights, Strasbourg, France
8. Kidwell LA, Fisher DG, Braun RL, Swanson DL (2013) Developing learning objectives for accounting ethics using Bloom's taxonomy. Acc Educ 22(1):44–65. https://doi.org/10.1080/09639284.2012.698478
9. Meeker QW, Hong Y (2014) Reliability meets BIg data: opportunities and challenges. Qual Eng 26(1):102–116
10. Mittelstadt BD, Allo P, Taddeo M, Wachter S, Floridi L (2016) The ethics of algorithms: mapping the debate. Big Data Soc 3(2):1–21
11. Nentl N, Zietlow R (2008) Using Bloom's taxonomy to teach critical thinking skills to business students. Coll Undergrad Lib 15(1–2):159–172. https://doi.org/10.1080/10691310802177135
12. Ozdemir V, Hekim N (2018) Birth of industry 5.0: making sense of bog data with artificial intelligence, "the internet of things" and next generation technology policy. OMICS J Integrat Biol 2018:65–76. https://doi.org/10.1089/omi.2017.0194
13. Parry O, Mauthner NS (2004) Whose data are they anyway?: practical, legal and ethical issues in archiving qualitative research data. Sociology 38(1):139–152
14. Ryan M, Antoniou J, Brooks L, Jiya T, Macnish K, Stahl B (2019) Technofixing the future: ethical side effects of using AI and big data to meet the SDGs, 2019 IEEE SmartWorld. In: Ubiquitous intelligence & computing, advanced & trusted computing, scalable computing & communications, cloud & big data computing, internet of people and smart city innovation, Leicester, United Kingdom, pp 335–341. https://doi.org/10.1109/SmartWorld-UIC-ATC-SCALCOM-IOP-SCI.2019.00101

15. Schradie J (2017) Big data is too small: research implications of class inequality for online data collection. Media and class: TV, film and digital culture. Edited by June Deery and Andrea Press. Abingdon, UK: Taylor & Francis

16. Sims RR, Felton EL Jr (2005) Successfully teaching ethics for effective learning. College Teach Meth Styles J Third Quarter 1(3):31–48

17. Topol E (2019) High-performance medicine: the convergence of human and artificial intelligence. Nat Med 25(2019):44–56

18. Wolf B (2015) Burkhardt Wolf: big data, small freedom? / radical philosophy. Radic Philos. https://www.radicalphilosophy.com/commentary/big-data-small-freedom. Accessed 13 May 2019

Chapter 7
Planning for 5G: A Network Perspective on Quality of Experience

7.1 Introduction

The 5G technology is designed to replace 4G technology in order to better support more demanding services and applications. The design of 5G communication networks is mainly based on virtualisation and the network can be abstractly viewed as one unified hardware platform, based on which, a number of very diverse services can be provided. In essence, 5G architecture is built around service delivery.

As services and applications become the major concern for 5G network providers, it follows that the users subscribing to these services should be happy with the network performance in the form of end-to-end quality of experience. The end-to-end approach, which goes back to viewing the network as a unified platform, is also the key to how the 5G communication networks address the issue of user quality of experience.

Whereas in 4G networks, all users are in contention for the same set of services; in the case of 5G, with an increased and very diverse set of services, the network resources can be divided between different groups of services that have similar characteristics. This is referred to as network slicing and it takes place end-to-end, i.e. we can have slices in each part of the network. End-to-end network slicing is, arguably, the defining 5G feature.

The concept of slicing is not new. In fact, it has been partially implemented in mobile networks of previous generations, as domain-specific (e.g. core) slicing. However, in the 5G network, slicing splits resources into logical or virtual networks ("slices") across domains, to address use cases with distinct characteristics and service level agreement (SLA) requirements [7]. The network slicing mechanism is one of the mechanisms used by 5G to ensure end-to-end service delivery and service quality that ultimately guarantees user quality of experience. The purpose of network slicing is thus to enhance quality of experience of 5G users.

© Springer Nature Switzerland AG 2021
J. Antoniou, *Quality of Experience and Learning in Information Systems*,
EAI/Springer Innovations in Communication and Computing,
https://doi.org/10.1007/978-3-030-52559-0_7

7.2 Network Slicing

The need for network slicing originates from the fact that 5G communication networks introduce many new and diverse services. Some of these services include smart homes, smart wearable devices, automated vehicles and many others. As a consequence, a variety of business models needs to also be supported, including operator-provider services, but also third-party service provisions, as well as other business models. Looking back at how 4G networks handled services, in a "one-size fits all manner", the designers of 5G came to the conclusion that such type of solution is not appropriate for a 5G communication network. A solution that can optimise performance, and consequently QoE, for the different sets of service and application requirements would be more successful.

The idea of network slicing is about dividing the one physical network end-to-end into a number of virtual slices, such that each slice can be managed independently, by having dedicated resources that are not used by other slices. This is done so that each slice can be optimised separately to meet the diverse sets of 5G service requirements.

7.3 Network Slicing and 5G Service Requirements

To get an idea of these requirements, we can explore the ITU definition for 5G that defines a set of targets for 5G communication networks. With regards to the peak data rates provided for users, the downlink target is 20 Gbps and the uplink target is 10 Gbps, whereas the cell edge data rate should be targeted at 100 Mbps for the downlink and 50 Mbps for the uplink. With regards to the latency experienced in the radio access network, the target is to stay below 1 ms. With regards to the target connection density, 5G should support 1 million devices per squared km, and the cell throughput density should be 10 Mbps for every squared meter [2].

It is clear that with these target service demands are quite high and it is important to find a smart solution for resource management in 5G to target satisfactory user quality of experience. Network slicing provides such solution. Hence, ITU has proposed three (3) usage scenarios that can serve as corresponding standard network slices, with each targeting different service requirements. These scenarios are also referred to as Slice Service Types. Within the context of 5G network slicing, a Slice Service Type defines the expected behaviour of a network slice in terms of specific features and services. The three standardised Slice Service Types are:

- The *eMBB*, which stands for *enhanced Mobile Broadband*
- The *URLLC*, which stands for *Ultra Reliable Low Latency Communications*
- The *mMTC*, which stands for *Massive Machine Type Communications*

The first Slice Service Type, referred to as enhanced Mobile Broadband scenario, targets very high data rate applications such as ultra-high-definition video and voice as well as 3D applications. The main requirement addressed by this scenario is to achieve high peak data rates, specifically 20 Gbps for the downlink and 10 Gbps for the uplink.

The second Slice Service Type, referred to as Ultra Reliable and Low Latency Communications scenario, targets applications such as augmented reality, self-driving cars, critical healthcare applications and other applications. The main requirement addressed by this scenario is achieving very low latency in the radio access network, specifically less than 1 ms.

The third Slice Service Type, addressed by a third ITU scenario, addresses applications that are based on high device density such as smart cities and smart homes. It is referred to as the Massive Machine Type Communications scenario and deals the connection density requirement of 1 million devices per squared km, focusing on IoT types of devices.

Given that network slicing is a wanted solution to support user quality of experience in 5G networks, the amount of slices needs to be limited in order to avoid a resource scarcity issue. This is the reason that each of the limited number of slices needs to be defined for *groups* of applications with similar resource requirements and corresponding quality requirements, and not for single services or applications. The additional challenge that arises from this grouping of services is that in such a scenario, each service or application has its own specific requirements in addition to the main target requirement for the group, which characterises any specific slice.

7.4   Single Slice Service Type and Resource Management

Therefore, the specific service requirements and hence the specific quality within a single slice needs to be preserved between services and applications themselves so that user quality of experience is optimised, no matter which of the applications within a slice the user subscribes to. Additional 5G capabilities such as Software Defined Radio and Network Functions Virtualisation could support this effort [1].

When network slices are designed, it is important to pay attention to ensuring a consistent high level of QoE for the 5G user. While there may be fluctuations in network quality and performance, as well disruptions and unpredictable interference, these should be kept at minimal level, such that each service is delivered to the 5G user, in a satisfactory manner that hides the complexity of this effort, i.e. the use of slices, etc. To achieve at least a satisfactory experience, 5G networks need to ensure an efficient delivery of remote services and data to end-users.

Often this will be achieved through making use of cloud data centres hosted by the cloud provider's infrastructure. Nevertheless, different types of 5G applications will need different QoE requirements, and thus, using a personalised QoE management solution is somehow expected in a 5G ecosystem. Cloud computing and corresponding real-time computations and large-scale online modelling can support such personalised user QoE requirements [8].

7.5   A Simple Resource Management Scenario

We may have a closer look at such a resource management scenario within a specific Slice Service Type. The approach that we will be following in order to understand how such resource management will work is a modelling approach. We will approach this from a game theoretic perspective by looking at a simplified network slice.

Let us consider that two high data rate applications are deployed within the same Slice Service Type, specifically the enhanced Mobile Broadband (eMBB). Applications assigned to the eMBB have resources committed or reserved for them in order to achieve the high bandwidth demands. We are considering the existence of two application with very high bandwidth demands in this scenario, but, we assume that each of the applications has a different latency requirement; one has a maximum latency requirement of t milliseconds (ms), which needs to be guaranteed, and the second has a more flexible latency requirement, i.e. which can exceed t ms.

What happens once a user tries to use these two applications depends on which application the user chooses to use. Based on the choice of the application, resource management mechanisms of the particular slice will schedule the use the network resources accordingly.

Let the amount of network resources, e.g. bandwidth, needed to guarantee a latency of t ms, be equal to r. Remember that this scenario is already equipped to satisfy large bandwidth demands for the group of applications it supports, so the strategy of guaranteeing more bandwidth resources to ensure low latency is a good strategy.

For the first application, i.e. the application with strict latency demands, the time cost of using the network must always be t ms, so r amount of resources are reserved for each user that chooses to use this application (or that subscribes to it). Given that this is a requirement, the required latency must be guaranteed by the network resource management mechanisms, no matter of the amount of network traffic.

For the second application, the time cost of using the network varies according to the number of 5G users that use the network resources controlled by the specific slice. The more the users on the network, the less the resources are available to be divided amongst them. Also, the less the resources are, the higher the latency becomes. Assuming that network resources for the second application are limited at any given time based on the number of users subscribed for the first application, then, the number of users that use the second application is the only factor that affects the fraction of the remaining resources that each user of the second application is assigned. That in turn, affects the time cost of the connection and hence the latency that each user experiences.

Assume that the network needs to reserve the resources for the users of the first application. To achieve that, the users must subscribe to the service, so that the network resource management can set aside enough resources for these users. Let $R = r_1 \cdot n$, where r_1 is the amount of resources needed per user of the first application, and n is the number of subscribed users for the first application. The second application can use the remaining resources B, where $B = A - R$, and A is the total available resources for the slice. Remember that R is the amount of resources needed for the first application as calculated above. Let there be m users for the second application. Then each of those users will be assigned r_2 amount of resources, where $r_2 = B/m$.

Thus, consider that the resources assigned per user of the second application is a linear function f such that $f(x) = x$, where x is the fraction of users using the second application, and $x \in [0,1]$. Let the maximum amount of resources assigned to a particular user be equal to r_1, i.e. the amount of resources needed to ensure the required latency for the first application. Given a total of n users that are subscribed for the first application, then the amount of resources that must be reserved within the slice is $R = r_1 \cdot n$.

We will make a side note here. Several network providers only reserve a fraction of the total necessary resources for their subscribed users, given the probability of simultaneous need for these resources, such that instead of R, the amount reserved is equal to $p \cdot R$, where $p \in [0,1]$. It is not important in this case, because what we care about is that the amount of resources reserved for the subscribed users of the first application is a constant and will remain the same constant, no matter how many of the users are actually using the application at the same time. For simplicity purposes, we will simply use the term R to refer to this amount of resources.

Another important assumption for this example is the following: if a user subscribed to the first application is not using that application, then that specific user may decide to use the second application, which does not require subscription. Thus, given the function for the usage of resources for the second application, which is defined as $f(x) = x$, if all users that are using the network slice at any given time decide to use the second application, then the fraction of users x, $x \in [0,1]$ will be equal to 1. That translates into using the total amount of available resources for the second application.

We can proceed to make a preliminary conclusion at this point. For the user experience to be ensured, then the amount of resources available to serve users of the second application should be such that if all users decided to use the second application at the same time, then that is possible. Therefore, r_2, i.e. the amount of resources allocated to a single user for using the second application should only be as small as the minimum amount of resources necessary to ensure minimum acceptable application quality or minimum QoE. By looking at the given function, also known as identity function, i.e. $f(x) = x$, we observe that all users at any time could potentially be users of the second application.

This is a simplistic function, but even if we generalise the function to $f(x) = ax + b$, where $a, b \geq 0$, then the result will still vary according to the number of users. We will return to looking at what this function means later in the chapter.

For now, we will investigate to the more simplistic model of using the identity function to analyse the resource management scenario. Let us try to analyse the best possible solution in terms of finding the best solution for the network resources, as well as trying to satisfy the required amount of resources per user, for selfishly acting network users.

7.6　System Efficiency and Price of Anarchy

When we consider how well a system works, we often use the measure for the system's *Price of Anarchy* [3]. This measure considers two numerical values for the model combined together. On the one hand, it considers the best solution from the user's perspective, i.e. in the case that all users acted selfishly, i.e. selected their dominant strategy, where this exists. On the other hand, the measure of the Price of Anarchy considers the best solution that would be given for the system by a centralised authority, if this were possible, so that users are controlled such that the best usage of the network is made. If these two numbers are close together, i.e. if the Price of Anarchy, which is the ratio between these two numbers, is close to 1, then the system design is considered efficient. This is the required outcome for our resource management scenario, because an efficient system design can guarantee an increased user QoE.

Before returning to the analysis of our model, we will briefly introduce the Price of Anarchy metric for systems. The term Price of Anarchy was first used by Elias Koutsoupias and Christos Papadimitriou in 2009 [3], but the idea of measuring inefficiency of equilibrium was explored before their publication. The concept is now used to study systems efficiency, especially systems that can be modelled using the tools of algorithmic game theory.

It is important to understand the significance of this measure because it juxtaposes the users' selfish behaviour and the most efficient system solution. It can, therefore, show how much a system performance or efficiency degrades when the users of a given system act selfishly. This is mostly important when we evaluate systems that exist and we cannot change them, e.g. systems that exist in nature or a fixed infrastructure. When designing a system especially when it comes to technology, algorithmic design or mechanism design is an important step. However, for systems that already exist, the Price of Anarchy can be a very useful measure.

Considering the example we are presenting of resource management for users of two different services of the same 5G slice, the interpretation of the Price of Anarchy can be as follows. When we refer to system efficiency, we can interpret it to mean the efficient use of system resources such that the system resource availability is not strained by user demands. Alternatively, selfish behaviour could imply the best use of resources per user such that desirable QoE is achieved. Thus, the Price of Anarchy measures the ratio between users' service request according to users' best selfish behaviour and the best service allocation to users according to most efficient system use, at any given time.

Price of Anarchy is a good measure for system efficiency that provides the designers of a system with more information than simply finding the system equilibrium. To find the Price of Anarchy, we need to model the system as a game and allow the players to figure out the best, i.e. selfish strategies. The efficiency of the system is also calculated as a function of the system, given the parameters of the system, e.g. latency, available bandwidth, etc. It is often possible to model selfish behaviour in a game by using the concept of equilibriums, i.e. the Nash equilibrium, that given the best action of each player when all other players have played their best moves. So, in essence, the Nash equilibrium is a concept used in game theory that results in a set of strategies, one for each player, such that no player has an incentive to change this strategy [6]. Different variations of the Nash equilibrium can lead to different variations of the Price of Anarchy measure, e.g. a Mixed Nash equilibrium can lead to a Mixed Price of Anarchy. However, in our example, we will be looking at a Pure Nash equilibrium, and hence, a Pure Price of Anarchy.

7.7 Efficiency and Equilibrium in the Network Slice Example

Returning to our example, we can assume that if a central network authority could decide the amount of users that would at any time request usage of either of the two available applications, then system efficiency could be achieved. On the contrary, as users decide on their actions, then each user would tend to act selfishly. In such a case, we need to consider all users' actions to determine a state that all users are satisfied, and investigate whether there is a Price of Anarchy such that the network traffic is manipulated somehow to show optimal behaviour. To do that, we will explore a known network model referred to as Pigou's example. This is usually used for congestion or scheduling examples in network like situations, so we will first describe the example itself and then try to see how this may apply in our situation.

Let us try to describe the simple network example known as Pigou's example: consider a double path parallel network with one source and one destination, the same for both paths. Thus, we may visualise two arcs connecting a source node to a destination node. However, each one of the two paths have different costs, e.g. taking each of the paths results in different travel times, because of different congestion and different characteristics of the path itself.

To define a Pigou-like network, consider a graph with two nodes and two links as shown in Fig. 7.1. The node s is the source of the network traffic and the node d is the destination of the network traffic. The variable x signifies the fraction of traffic that chooses each path, and κ is a constant value that remains unaltered by the fraction of traffic that decides to use the path.

Fig. 7.1 Visualising a Pigou-like network (for linear traffic)

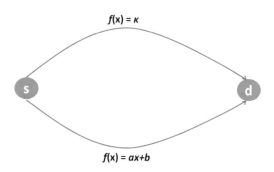

To help us model our example as a Pigou-like network, in order to proceed to define its Price of Anarchy, we first need to consider the definition of a Pigou-like network [5] and try to model our example according to the identified characteristics. A Pigou-like network can be defined as follows:

- A network with two nodes, a source and a destination.
- A network with two links, representing two paths from source to destination.
- The network can be used by the traffic entities, resulting in a traffic rate which is positive; we can represent the rate of traffic as x.
- One of the paths has a cost function; this can be anything, but for this example, we will consider linear functions, degree 1 polynomials, such that $f(x) = ax + b$.
- The second link has a constant cost, κ, which is equal to a cost equal to $f(x)$ for the first link evaluated at full cost; for example, if a equals to 1 and b equals to *zero*, we are left with the identity function $f(x) = x$. If we consider x to be a fraction of the traffic, such that $x \in [0,1]$, then the function $f(x) = x$ evaluated at full cost would be equal to 1, and that would be the constant cost of the second link, i.e. $f(x) = \kappa = 1$.

7.8   Calculating the Price of Anarchy

Let us consider the example assumption above, where the cost of the first link is $f(x) = x$, and the cost of the second link turns out to be equal to 1 per user, no matter what fraction of traffic uses it. In turn, we will try to decode users' desirable strategies. Given the user's choice of cost being equal to 1 for the one path, or, for the second path the cost being less than or equal to 1 (only in the case that all users choose the variable cost link), then the dominant strategy will be to take the link with the variable cost. This will be the dominant strategy for all users, so in the end one hundred percent (100%) of the users will choose this path and have on average a cost of $f(x) = \kappa = 1$.

It appears that the efficient strategy would be to split the users so that some of the users would take each path and the average cost would be less than 1. It turns out that by allowing half of the users to take each path, then half of them would experience a cost of 1 and half of them would experience a cost of 0.5, on average a cost of $\dfrac{1+0.5}{2} = 0.75$, which is less than 1, the cost calculated for the equilibrium solution.

To calculate the Price of Anarchy for the Pigou-like network, we would need to divide the equilibrium utility by the efficient system utility, i.e. $\dfrac{1}{0.75} = 1.333 = \dfrac{4}{3}$.

Pigou's example is often used to introduce some of the challenges of decision-making with regard to scheduling problems or traffic congestion problems in a situation that can be modelled as a network and reduced to a network of two nodes and two paths. In the specific scenario we are examining, we have one environment, the

5G slice, which is managed by a set of resource management mechanisms, and we have two options for a set of subscribed users, to select one of two applications at any given time. We hold that the selection of an application defines the specific path or option; however, the Pigou-like network model is still helpful to allow us to better understand the specific network resource management functionality. To examine this example similarly to a Pigou-like network, we must consider the following:

(a) It can be reduced to an environment with two states of being for the users, i.e. the state before selecting an application, and, the state after selecting an application, similarly to a source and a destination node to a network graph.
(b) There are two options for the users, to either choose the first or the second application, similarly to having two paths to follow to the destination node.
(c) The users are a non-negative number, which is a requirement for the number of units of traffic flow in a Pigou-like network.
(d) One of the application options needs resource guarantees so that the cost per user of using that link is always fixed, no matter the number of subscribed users, similarly to the link of fixed cost in the Pigou-like network.
(e) The second of the two paths has a cost that depends on the amount of traffic using the network and this varies. All of the subscribed users have the option to also use the second application. Therefore, the cost will vary according to a function of the amount of users choosing the second application at any given time, since there is a finite amount of resources available for the second application and each of the users only needs to have minimum quality requirements satisfied.
(f) The final requirement of a Pigou-like network is that the fixed cost of the first option should be equal to the maximum possible cost of the second option. This can be assumed for our example because the first application has the more demanding latency requirement and should always be guaranteed the maximum amount of resources.

We quantify the cost of using a particular option by the resource usage functions. If a fraction x_1 of the users selects the first application, then $f(x_1)$ represents the resources used to ensure the latency requirements for the first application; similarly, if a fraction x_2 of the users selects the second application, then $f(x_2)$ represents the resources used for the minimum latency requirements. The number of users choosing each application at a given time only affects the users choosing the second application because the users that choose the first application have a fixed network cost per user.

Then, by considering the Pigou-like network results that we explored previously, we can make some conclusions for this resource management problem. Let's say that a constant of 1, i.e. $f(x_1) = 1$, is the amount of resources that can be engaged per user to achieve the minimum requirements for the first application. This is also equal to the maximum amount of resources needed per user for achieving minimum quality requirements, as the increase in traffic congestion would require a higher amount of resources needed per user to achieve the necessary quality conditions. However, as the traffic congestion in the network decreases (or the specific slice of the network controlling quality end-to-end), then minimum requirements for the second application can be achieved with less resources. We will use the identity function for simplicity. Thus, the cost of a user choosing the second application is $f(x_2) = x_2$.

The monetary cost for the users is directly linked with the amount of resources engaged to serve the selected application. Therefore, given a choice, the users would select to use the second application instead of the first when possible to avoid cost. If at any given time, the users can all choose the second application, then they will do that because of the lower cost. Of course, this is not efficient for the network, because it engages more resources per user eventually. As the example is using the same functions as they are used in the Pigou-like network discussion, then the dominant strategy would result in the constant cost of 1. Additionally, a split of users such that fifty percent (50%) use the first application and fifty percent (50%) use the second application, will give the best use of resources, which result in an amount of 0.75, or three fourths of the maximum usage, per user.

Similarly, the Price of Anarchy, which gives us a measure of how selfish behaviour affects the system efficiency, remains the same. A Price of Anarchy equal to $\frac{1}{0.75} = 1.333 = \frac{4}{3}$ is close to one unit, and that is a satisfactory measure for system efficiency. What we learn is that the system behaves better if efficiently controlled centrally by the network design and not left to be the result of users' selfish decision-making.

Generalising: x_1 of the users will experience a resource cost of 1; x_2 of the users will experience a resource cost of x_2; $x_2 = 1 - x_1$, therefore, $(1 - x_1)$ of the traffic will experience a latency of x_2, i.e. $(1 - x_1)$.

The average latency of the system is thus: $x_1 + (1 - x_1)(1 - x_1) = x_1 + (1 - x_1)^2$. The optimal flow is thus the differentiation of this curve, which turns out to be 0.5. We already know this result from the identity function, but now we can consider that any linear degree 1 polynomial will differentiate to a constant fraction, with 0.5 being the best case scenario. Therefore, the optimal flow is for x_1 to be equal to 0.5 and thus x_2 will also be 0.5. If the traffic is split halfway through both paths, then the resulting average latency would be: $0.5 \cdot 1 + 0.5 \cdot 0.5 = 0.75$.

7.9   Discussion

The chapter discussed the importance of the Price of Anarchy for quantifying system efficiency. We have used a 5G example to show that system efficiency can be negatively affected by selfish behaviour for a simple example of sharing a 5G slice between applications, with only two options, i.e. two applications competing for resources. This is important for the overall study of user experience and QoE. In fact, designing or experiencing an inefficient network results in unsatisfactory QoE for the network users.

It is interesting that for the example in this chapter, in order for the QoE to be satisfactory, the users' preferences must be considered in conjunction with the system efficiency, which can also be viewed as the users' *social good*. Therefore, the efficiency metric eventually works out to the users' advantage for the specific example.

However, we have only considered an example of two options and we have mapped this on a network with two paths. It is expected that the reader will wonder about scenarios that can be mapped onto more complicated network graphs. A very interesting result was presented by Roughgarden [4] under the title "*The price of anarchy is independent of the network topology*". In essence, the results of this study show that under some conditions, system efficiency is not affected by a simple or complex network representation. The author studies the Price of Anarchy in an attempt to understand the effect of selfish user behaviour on overall network efficiency or performance. The game model studied considers non-cooperative users. The example used by the author is that of routing when selfish players make the decision of which path to follow. The cost measure is the delay or latency experienced by users on each link of the selected route. The delay is affected by the congestion on each link for some of the links, and there are the defined links with a fixed delay cost. The publication shows that selfish routing, i.e. allowing each user to select their preferred path (which results in an equilibrium strategy) does not minimise the total cost per user. However, it does decrease network performance.

Since the above results are shown based on a network's Price of Anarchy measure, the publication proceeds to show that this measure is, in fact, determined by the simplest networks, since:

> under weak hypotheses on the class of allowable edge latency functions, the worst-case ratio between the total latency of a Nash equilibrium and of a minimum-latency routing for any multicommodity flow network is achieved by a single commodity instance on a network of parallel links. [4]

Furthermore, under some cases of constant cost, then this simple two-node network of parallel links is enough to show the worst-case scenario, but we will not consider functions resulting in the worst case of the Price of Anarchy measure in this chapter.

Finally, what we would like to keep from this example is that ultimately, the user experience and overall QoE is affected by the technology. System efficiency needs to be a factor when evaluating user QoE. System efficiency prescribes the social good, which often needs to be considered in order to satisfy the user QoE.

As a final note in this chapter, which is the final in this book, we come around to the initial discussion, which led us to the exploration of different considerations of user experience. In order to design a truly user-centred digital product, there needs to be a consideration of both user QoE and user UX. There needs to be an evaluation of the user experience by looking at the environment, the system and the network efficiency, but also, by looking at the interface aesthetics, the user behaviour and the ease of learning that can take place.

References

1. Barakabitze AA et al (2020) 5G network slicing using SDN and NFV: a survey of taxonomy, architectures and future challenges. Comput Netw 167:106984, ISSN 1389-1286. https://doi. org/10.1016/j.comnet.2019.106984
2. International Telecommunication Union (2017) Radiocommunication sector – minimum requirements related to technical performance for IMT-2020 radio interface(s). Report ITU-R, M.2410-0 (11/2017), M Series: Mobile, radiodetermination, amateur and related satellite services
3. Koutsoupias E, Papadimitriou C (2009) Worst-case equilibria. Comput Sci Rev 3(2):65–69
4. Roughgarden T (2003) The price of anarchy is independent of the network topology. J Comput Syst Sci 67(2):341–364
5. Roughgarden T (2013) Selfish routing and the price of anarchy. In: Twenty lectures on algorithmic game theory. Cambridge University Press, Cambridge, pp 145–158
6. Spaniel W (2013) Game theory 101: the complete textbook. Create Space Independent Publishing Platform, San Bernardino. ISBN-13: 978-1492728153
7. Stavropoulos K (2019) 5G network slicing: what, why, what, how [Online]: www.exfo.com, published on Dec 2019. Accessed in April 2020
8. Wang Y, Li P, Jiao L, Su Z, Cheng N, Shen XS, Zhang P (2016) A data-driven architecture for personalized QoE management in 5G wireless networks. IEEE Wirel Commun Mag 24(1):102–110

Index

© Springer Nature Switzerland AG 2021
J. Antoniou, *Quality of Experience and Learning in Information Systems*,
EAI/Springer Innovations in Communication and Computing,
https://doi.org/10.1007/978-3-030-52559-0

Printed in the United States
by Baker & Taylor Publisher Services